Wesley Mills

Outlines of lectures on physiology

With an introductory chapter on general biology

Wesley Mills

Outlines of lectures on physiology
With an introductory chapter on general biology

ISBN/EAN: 9783337217037

Printed in Europe, USA, Canada, Australia, Japan

Cover: Foto ©berggeist007 / pixelio.de

More available books at **www.hansebooks.com**

OUTLINES

OF

LECTURES ON PHYSIOLOGY,

WITH AN

INTRODUCTORY CHAPTER ON GENERAL BIOLOGY,

AND AN

APPENDIX CONTAINING LABORATORY EXERCISES IN
PRACTICAL PHYSIOLOGY,

BY

T. WESLEY MILLS, M.A., M.D., L.R.C.P., Eng.,

PROFESSOR OF PHYSIOLOGY, MCGILL UNIVERSITY,
MONTREAL.

MONTREAL:
W. DRYSDALE & CO.
1886.

TO THE

Present Students of McGill University,

AND TO THOSE WHO,

DURING THE PAST FIVE YEARS,

Have honored me with their confidence and respect

THIS LITTLE BOOK IS

𝔇𝔢𝔡𝔦𝔠𝔞𝔱𝔢𝔡.

I. Explanatory and Biological.

SCIENCE is exact correlated knowledge.

BIOLOGY is that science which treats of organized matter in general, whether animal or vegetable. It includes *Morphology* and *Physiology*.

MORPHOLOGY is the comparison of the *structure* of organisms with a view of ascertaining their relations.

PHYSIOLOGY deals with the *functions* of living things. It may be divided into *animal*, including human and comparative, and *vegetable* physiology.

BIOLOGY is concerned with *living* matter.

CHARACTERISTICS OF LIVING ORGANISMS.

Digestion and assimilation leading to growth, development, reproduction and death.

The physical basis of life is *protoplasm*.

Protoplasm is composed of C, H, O, N, with traces of P and S; small quantities of inorganic salts, especially phosphates; water in large proportion; and small quantities of proteids, carbohydrates and fats.

All organized beings have within themselves the energy to construct protoplasm.

Living matter can proceed only from living matter. The opposite view is upheld by the doctrine of *spontaneous generation*.

SYNOPTICAL STATEMENT OF THE CHARACTERS OF THE ORGANIC KINGDOM. (Pye-Smith.)

1. *Chemical:*
   Few elements; complex combination.
   Carbon compounds.
   Colloid condition; abundant water, rounded forms.

2. *Structural:*
   Cells and their derivatives: differentiated tissues, neither amorphous nor crystalline.
   Symmetry: spiral, radial, bilateral, serial.

3. *Functional:*
   Cycle of changes.
   Origin, from a parent organism; growth; assimilation; decay; death.
   Irritability; movement; heat.
   Reproduction.

The relations of Plants and Animals may be thus expressed (modified from Brass.)

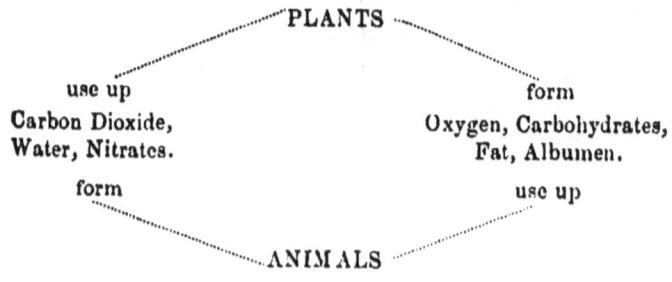

DISTINCTIVE CHARACTERS OF ANIMALS AND PLANTS.

(After Pye-Smith, modified.)

1. Chemical.

Animals.	Plants.
Predominance of N compounds.	Predominance of Starchy compounds.
	Cellulose wall to each cell.
Presence of hæmoglobin.	Presence of chlorophyll.

2. Structural.

Digestive cavity, nervous and muscular tissues.	

3. Functional.

Animals	Plants
Nutrition from organic food only.	Nutrition from inorganic food. Fixation of C.
Functions of relation predominant.	

Exceptions:

1. Fungi and bacteria have no chlorophyll and feed on organic food.

2. Have a large proportion of N in their composition; are deficient in carbohydrates.

Carnivorous plants (pitcher plant, sundew and Venus' fly-trap), live partly on organic food.

Certain animals, *e.g. Hydra viridis*, contain chlorophyll.

Ascidians form cellulose; *Ascidians* and some *worms* form starchy matters. Most animals with livers form animal starch (glycogen). It is, in the present state of knowledge, impossible to give characteristics by which plants and animals may in *all* cases be distinguished.

GENERAL BIOLOGY.

Many of the statements of the foregoing chapter may be verified by the study of the following examples:

PLANT FORMS.

I. YEAST (*Torula or Saccharomyces Cerevisiæ*).

1. *Morphological:*
   - (a) One-celled plants.
   - (b) Protoplasmic, clear, semi-fluid mass surrounded by a *cellulose* wall. Protoplasm may contain *vacuoles*.
   - (c) Propagated by *budding* and by *endogenous* division.
   - (d) Contains no chlorophyll.

2. *Functional and Chemical:*
   Causes *fermentation* in saccharine substances; the products being alcohol and carbon dioxide.

$$C_6 H_{12} O_6 = 2\,CO_2 + 2\,C_2 H_5.OH.$$

Torula consists of:

1. A proteid resembling casein.
2. Cellulose.
3. Fat.
4. Water.

The *cell-wall* contains the whole of the *cellulose* and a little of the mineral matter.

The *protoplasm* contains the remaining substances and most of the inorganic salts.

In ultimate analysis the above constituents yield the following chemical *elements*: C, H, O, N, S, P, K, Mg, Ca; but the five first in much larger proportion than the others.

Yeast, when burned, leaves an *ash* containing the above elements.

The rapid multiplication of the yeast cells implies the power to construct protoplasm, etc., from the fluid in which they live, their food.

An artificial food may be made of the following composition (Pasteur's solution):

Water, $H_2 O$.
Cane Sugar $C_{12} H_{22} O_{11}$.
Ammonium Tartrate $C_4 H_4 (NH_4)_2 O_6$.
Calcium Phosphate $Ca_3 (PO_4)_2$.
Potassium Phosphate $K H_2 P O_4$.
Magnesium Sulphate $Mg S O_4$.

Torula does not require *free* O, but can derive the necessary supply from its food (sugar).

It can also thrive in fluids in which ammonium tartrate is replaced by pepsin and sugar.

The life and fermentative action of torula is dependent upon a certain range of temperature.

Experiments demonstrating the above properties.

II. BACTERIA.

Morphological:

    1. Of various forms: globular, oblong, rod-like, spiral, etc.

    2. Multiply by transverse division.

    3. Consist of protoplasm *without* chlorophyll.

    4. Have a mobile and resting stage (Zoogloea-stage).

    5. Grow and reproduce in Pasteur's and similar fluids.

    6. Are killed by certain temperatures; but their *spores* (reproductive forms) resist a much higher temperature than the parent forms.

    7. All kinds of putrefaction are dependent upon their presence.

    8. Like the yeast-cell they bear drying and are readily transported through the air.

    Principal forms; (a) *Micrococcus* (b) *Bacillus*, (c) *Vibrio*, (d) *Spirillum*, (e) *Spirochæte.*

Demonstration: Treatment of hay infusion, urine or other fluid suitable for their growth, to illustrate the above laws.

III. MOULDS OR FUNGI.

I. PENICILLIUM GLAUCUM. (Ordinary Green Mould.)

1. Consists of interlacing *filaments* constituting a felt work (*mycelium*.)
2. Each filament is made up of protoplasm without chlorophyll, and a cellulose wall.
3. Those filaments projecting into the air constitute *aerial hyphæ*; those descending into the fluid or beneath the general level are the *submerged hyphæ*.
4. The filaments have transverse partitions at intervals; some end in rounded extremities; others in a kind of branches (*conidiophores*) which, dividing transversely, form *spores*.
5. The spores are of essentially the same structure as torula, and are denominated *conidia*.
6. In suitable food-material the conidia grow into filaments.

II. MUCOR MUCEDO. (White Mould.)

In structure it is very similar to *penicillium*. Its mode of reproduction is somewhat different.

1. *Asexual:* A branch from a hypha after attaining a certain length dilates into a rounded head or *sporangium*. Its protoplasmic contents divide and the subdivisions become coated with a cellulose wall constituting *spores*. The wall of the sporangium, getting gradually thinner, bursts, setting free the contents known as *ascospores*, while the case or sporangium is an *ascus*.

This is the ordinary method of reproduction when the mould grows in fluid and nourishment is abundant.

2. *Sexual*: two *hyphae* with dilated extremities approach; the contents of the enlarged ends mingle into one mass (conjugation) known as a *zygospore*. The cellulose covering consists of an outer coat, known as an *exosporium*, and an inner, *endosporium*. Under favorable conditions the zygospore, finally, after bursting, germinates into a hypha ending in a *sporangium* which produces spores in the usual way. This entire process illustrates in the vegetable world *alternation of generations:*

Generation A gives rise to B, and B gives rise to A again; but mostly there are several generations of B before a return to A takes place.

Mucor when submerged in a saccharine liquid reproduces, by a process of *transverse division* into forms which become rounded and then bud as does torula. At the same time a *fermentation* is excited.

IV. PROTOCOCCUS (Protococcus Pluvialis).

Is a spheroidal *one-celled* plant consisting of a clear cellulose wall and granular protoplasmic contents. Throughout the latter a red or green *chlorophyll* is distributed.

The usual mode of reproduction is by *fission*; gemmation is rare.

Like all plants containing chlorophyll, *sunlight* is essential to its life.

Under the influence of sunlight CO_2 is decomposed; the C is "fixed" and the O given off.

The *fungi, torula* and the *bacteria* (in most cases) require (free) O and give off CO_2.

Protococcus flourishes in rain-water, therefore must construct protoplasm from what is found therein alone, i.e., CO_2, $(NH_4) NO_3$, small quantities of earthy salts, etc.

Protococcus is thus a *typical* green plant.

Motile form of protococcus: The protoplasmic contents protrude through the cell-wall in the form of vibratile *cilia*. The cell-wall may disappear, and the protoplasm may undergo division; it generally returns finally to the resting state.

Like the other forms of plants already considered it bears *drying* without loss of vitality, and can be readily transported in this condition by winds.

I. AMŒBA OR PROTEUS ANIMALCULE.

A one-celled typical animal.

It consists of a microscopic mass of jelly-like protoplasm in which the following parts may be distinguished: (*a*) *Ectosarc* firmer and clearer. (*b*) *Endosarc* more fluid and granular. (*c*) *Nucleus*. (*d*) *Nucleolus*. (*e*) Contractile *vacuole*. (*f*) *Pseudopodia*. There is no cell-wall. Amœba may exist in a less differentiated form. The amœba reproduces by (*a*) *fission*, (*b*) *gemmation*.

All the activities of amœba are permanently arrested at 0° C. and 4 5° C.

It has the following properties :
It is Contractile.
 Irritable and automatic.
 Receptive and assimilative.
 Metabolic and secretory.
 Respiratory.
 Reproductive.

The above properties belong to protoplasm in general, whether of plants or animals. It is clear that as *amœba* can discharge all the functions of the highest animal it may be considered the simplest *animal type*.

II. GENERAL MORPHOLOGY.

The units of which animals and plants are made up a *cells*.

A cell consists of a *wall*, *contents*, a *nucleus* and, sometimes a *nucleolus*.

The nucleolus and cell-wall may be wanting.

The *animal* cell is generally without cell-wall.

The *nucleus* seems to determine largely the behavior of the cell, especially in reproduction.

The *ovum* from the development of which the animal is derived is itself a cell.

At one period of the *development* of the most complex animal forms the organism is composed wholly of cells.

Later in most animals there are, in addition to cells

tissues which are made up of cells, modified cells and "formed" material or secreted material, all derived from cells.

Every cell arises from some antecedent cell by *fission*, *endogenous* multiplication, *gemmation*, etc.

The simplest animals and plants as amœba and protococcus are *unicellular* organisms.

Some animals remain but little more than *colonies* of amœbæ, as the sponges and some other cœlenterates.

The *Infusoria* may be regarded as modified amœbæ; they are *unicellular*. Animals may be divided in accordance with their structure into two great groups:

1. *Protozoa* (Protista), one-celled organisms of which *Amœba*, *Paramœcium* (slipper animalcule) and *Vorticella* (bell animalcule) are readily obtainable examples.

2. *Metazoa*, consisting of aggregations of cells and including all animals not belonging to the first group.

A knowledge of the structure prevailing in the lower divisions of the group metazoa, may be derived from a study of changes taking place in the *ovum* of many animals.

The ovum by *segmentation* divides into a number of cells which are finally grouped in a certain order, varying in different divisions of the Metazoa. Of special interest are the *morula* and the *gastrula* forms.

In the *morula* the segments form a mulberry-like spherical mass. The cells may so separate from one another as to give rise to a *segmentation* cavity within.

The cells may be arranged around the latter by *invagination* (inversion), or by *delamination* giving rise to the *gastrula*.

The segmentation cavity (*archenteron*) opens outward by the *blastopore*. The outer layer of cells is the *epiblast*; the inner, the *hypoblast*. These, with the intermediate *mesoblast*, constitute the "germ-layers" of the vertebrate ovum.

The *gastrula* form gives a good idea of the structure of *hydra*, the fresh-water polyp. The latter is composed of an outer layer of cells (*ectoderm*,) and inner layer (*entoderm*) which enclose a cavity answering to the stomach, etc.

The *tentacles* may be regarded as *tubular* expansions of the sac.

The *mouth* corresponds to the *blastopore* of the gastrula form.

Hydra reproduces by *gemmation*, but in the summer season *testes* form at the bases of the tentacles and *ovaries* at the attached end of the animal.

A polyp may be regarded as a *colony* of *amœbæ*, modified, and arranged as a double-walled sac.

Such structure is typical of the whole group *cœlenterata* (sponges, jelly fishes, hydroids, corals, etc.)

If a hydra be divided into segments, each of the latter can develop into an entire animal.

The *neuro-muscular* cells and the *urticating capsules* (*thread cells, nematocysts*) are of special interest.

II. Physiological. Introductory.

Physiology is the science which treats of the *functions* of living organisms.

Vitality and function are the resultant of certain forces, chemical and physical.

Life may be regarded as the sum or resultant of the energies of living matter.

Energy may be *potential* (latent) or *kinetic*.

Kinetic energy is evidenced in motion, whether of molecules and atoms or of masses.

One kind of energy is representable and exchangeable into another; but energy or force cannot be annihilated.

Consumption of energy; correlation of forces. Vital energy may be transformed chemical energy, latent or potential in food and rendered kinetic in the form of heat and motion.

Physiology is advanced only by observation and experiment.

It is an experimental science. "Experimental" is often used in a sense opposed to "chemical" physiology.

Modern physiology aims at ascertaining the chemistry and physics of *protoplasm*.

As a matter of fact, physiology has, hitherto, investigated the functions of certain well-known mammals (rabbit, dog, cat, and, to a less extent, the horse and a few others), chiefly.

Human physiology has been studied in a few cases of traumatic fistulæ, &c.; but a great part of the so-called "human" physiology is the product of inference and deduction from the results obtained by the experimental study of the lower animals and from clinical and pathological investigation. The latter must always be received with great caution in the realm of physiology. Its chief value lies in its *suggestiveness*.

The physiology of the invertebrates, and in great part of the lower vertebrates, is yet to be wrought out; but till such is accomplished there can be no complete human physiology.

In the advancement of the latter branch of the science much may be done by the individual in the form of observation and experiment on himself.

Modern physiology and psychology are closely related.

It will appear from the foregoing sections that the study of physiology implies the application of anatomy, chemistry and physics at every step.

The tissues are susceptible of the following *physiological* division:

  Contractile.
  Irritable and automatic.
  Secretory or excretory.
  Metabolic.
  Reproductive.
  Indifferent, mechanical or storage.

III. Chemical Constitution of the Body.

Such food as supplies energy directly must contain *carbon* compounds.

Living matter or protoplasm always contains *nitrogenous* carbon compounds.

In consequence C, H, O, N, are the elements found in greatest abundance in the body.

The elements S & P are associated with the nitrogenous carbon compounds; they also form metallic sulphates and phosphates.

Cl and F form salts with the alkaline metals Na, K and the earthy metals Ca and Mg.

Fe is found in *hæmoglobin* and its derivatives.

Protoplasm, when submitted to chemical examination, is killed. It is then found to consist of proteids, fats, carbohydrates, salines and extractives.

It is probable that when living it has a very complex molecule consisting of C, H, O, N, S and P chiefly.

PROXIMATE PRINCIPLES.

1. Organic
   - (a) Nitrogenous. { Proteids. Certain crystalline bodies.
   - (b) Non-nitrogenous. { Carbohydrates. Fats.

2. Inorganic { Mineral salts. Water.

SALTS.—In general the salts of *sodium* are more characteristic of *animal* tissues and those of *potassium* of *vegetable* tissues.

Na Cl is more abundant in the *fluids* of animals; K and phosphates more abundant in the *tissues*.

Earthy salts are most abundant in the harder tissues.

The salts are probably not much, if at all, changed in their passage through the body.

In some cases there is a change from acid to neutral or alkaline.

The salts are essential to preserve the balance of the nutritive processes. Their absence leads to disease, *e.g.*, scurvy.

GENERAL CHARACTERISTICS OF PROTEIDS.

They are the chief constituents of most living tissues, including blood and lymph.

The molecule consists of a great number of atoms (complex constitution), and is formed of the elements C, H, N, O, S and P

All proteids are amorphous.

All are non-diffusible, the *peptones* excepted.

They are soluble in strong acids and alkalies, with change of properties or constitution.

In general they are coagulated by alcohol, ether and heating.

Coagulated proteids are soluble only in strong acids and alkalies.

Classification and Distinguishing Characters of Proteids.

1. *Native albumins:* serum albumin; egg albumin; soluble in water.

2. *Derived Albumins (albuminates):* Acid and alkali albumin; casein; soluble in dilute acids and alkalies, insoluble in water. Not precipitated by boiling.

3. *Globulins:* globulin (globin); paraglobulin; myosin; fibrinogen. Soluble in dilute saline solutions and precipitated by stronger saline solutions.

4. *Peptones:* soluble in water; diffusible through animal membranes; not precipitated by acids, alkalies or heat. Derived from the digestion (peptic, pancreatic) of all proteids.

5. *Fibrin:* insoluble in water and dilute saline solutions. Soluble, but not readily, in strong saline solutions and in dilute acids and alkalies.

CERTAIN NON-CRYSTALLINE BODIES.

The following bodies are allied to proteids, but are not the equivalents of the latter in the food.

They are all composed of C, H, N, O. Chondrin, gelatin, keratin have in addition S.

Chondrin: the organic basis of cartilage. Its solutions set into a firm jelly on cooling.

Gelatin: the organic basis of bone, teeth, tendon, etc. Its solutions set (glue) on cooling.

Elastin: the basis of elastic tissue. Its solutions do not set jelly-like (gelatinize.)

Mucin: from the secretion of mucous membranes; precipitated by acetic acid and insoluble in excess.

Keratin: derived from hair, nails, epidermis, horn, feathers. Highly insoluble.

Nuclein: derived from the nuclei of cells. Not digested by pepsin; contains P but no S.

THE FATS.

The fats are hydrocarbons; are less oxidised than the carbohydrates; are inflammable; possess latent energy in a high degree.

Chemically, the neutral fats are glycerides or ethers of the fatty acids, *i.e.*, the acid radicles of the fatty acids of the *oleic* and *acetic* series replace the exchangeable atoms of H in the triatomic alcohol glycerine, *e.g.*

Glycerine. Palmitic Acid. Glycerine tri-palmitate or Palmitin.

$$C_3H_5 \begin{cases} OH \\ OH \\ OH \end{cases} + \begin{cases} HO.OC.C_{15}H_{31} \\ HO.OC.C_{15}H_{31} \\ HO.OC.C_{15}H_{31} \end{cases} = C_3H_5 \begin{cases} O.CO.C_{15}H_{31} \\ O.CO.C_{15}H_{31} \\ O.CO.C_{15}H_{31} \end{cases} + 3H_2O$$

A *soap* is formed by the action of caustic alkalies on fats, *e.g.*

Tripalmitin. Potassium Palmitate.

$$\begin{matrix} C_3H_5 \\ (C_{16}H_{31}O)_3 \end{matrix} \Big\} O_3 + 3(KOH) = 3 \left\{ \begin{pmatrix} C_{16}H_{31}O \\ K \end{pmatrix} O \right\} + \begin{matrix} C_3H_5 \\ H_3 \end{matrix} \Big\} O_3$$

The soap may be decomposed by a strong acid into a fatty acid and glycerine, *e.g.*

$C_{15} H_{31} . C O_2 K + H Cl = C_{15} H_{31} . CO_2 H + K Cl.$
Potassium Palmitate.   Palmitic Acid.

The *fats* are insoluble in water, but soluble in hot alcohol, ether, chloroform, etc.

The *alkaline* soaps are soluble in water.

Most animal fats are mixtures of several kinds in varying proportion; hence the melting point for the fat of each species of animal is different.

PECULIAR FATS.

Lecithin, Protagon, Cerebrin:

They consist of C, H, N, O, and the two first of P in addition.

They occur in the *nervous* tissues.

CARBOHYDRATES.

General formula $C_m (H_2 O)_n$.

1. THE SUGARS: *Dextrose* or grape sugar $C_6 H_{12} O_6 + H_2 O$ readily undergoes alcoholic fermentation; less readily lactic fermentation.

Lactose milk sugar $C_{12} H_{22} O_{11} + H_2 O$; susceptible of the lactic acid fermentation.

Inosit or muscle sugar $C_6 H_{12} O_6 + 2 H_2 O$; capable of the lactic fermentation.

Maltose $C_{12} H_{22} O_{11} + H_2 O$, capable of the alcoholic fermentation. The chief sugar of the digestive process.

All the above are much less sweet and soluble than ordinary cane sugar.

2. THE STARCHES:

Glycogen $C_6 H_{10} O_5$ convertible into dextrose. Occurs abundantly in many fœtal tissues and in the liver, especially of the adult animal.

Dextrin $C_6 H_{10} O_5$ convertible into dextrose. Soluble in water; intermediate between starch and dextrose; a product of digestion.

Pathological: grape sugar occurs in the urine in *Diabetes mellitus.*

Certain substances formed within the body may be regarded as, chiefly, waste-products, the result of metabolism or tissue changes.

They are divisible into nitrogenous metabolites and non-nitrogenous metabolites.

Nitrogenous Metabolites:

1. Urea, uric acid and compounds, kreatinin, xanthin, hypoxanthin (sarkin), hippuric acid, all occurring in urine.

2. Leucin, tyrosin, taurocholic and glycocholic acids, which occur in the digestive tract.

3. Kreatin, constantly found in muscle; and a few others of less constant occurrence.

The above consists of C, H, N, O. Taurocholic acid contains also S.

The molecule in most instances is complex.

Non-Nitrogenous Metabilites.

These occur in small quantity, and some of them are secreted in an altered form.

They include lactic and sarcolactic acid, oxalic acid, succinic acid, etc.

Demonstration: The properties of the bodies considered in this chapter.

IV. Blood and Lymph.

BLOOD.

Blood with Lymph constitutes the great "*internal medium*" of the tissues. (Bernard).

Blood is morphologically a *tissue* in which the plasma represents the matrix and the corpuscles the cell elements.

Living blood in the vessels consists of *plasma* and *corpuscles*; blood when shed rapidly dies resulting in *coagulation*.

Histological : 1. *Colorless corpuscles*: relative numbers, form, size, etc.

They constitute the sole morphological element in the blood of most invertebrates.

2. *Red Corpuscles*: size, shape, &c., in the different groups of vertebrates.

Their peculiarities in the *mammalia*.

The red corpuscles of man.

Nucleated in the fœtus.

Distinction of *stroma* and *hæmoglobin*.

The colorless corpuscles are morphologically and physiologically amœboid cells (amœbæ.)

Physiological:
Blood within an arteriole; within a capillary.
Blood as it issues from a divided artery; a divided vein.

Phenomenon of Coagulation: As seen in blood shed into a vessel. *Stages:* gelatinous stage; solid clot (coagulum); cupping; expression of serum.

Coagulation of a drop of freshly drawn blood, under the microscope; fibrin threads; *rouleaux.*

Clotting in a capillary tube.

Whipping or defibrinization of blood.

Physical condition of blood:

(a) Before coagulation as ⎰ Plasma.
    in a blood vessel : ⎱ Corpuscles.

(b) After clotting: Clot. ⎰ Fibrin.
    ⎱ Corpuscles.
    Serum.

(c) Defibrinated blood : ⎰ Corpuscles.
    ⎱ Serum.

Theories of Coagulation:

1. Coagulation is due to the action of a fibrin ferment on paraglobulin (fibrinoplastin) and fibrinogen (A. Schmidt).

Ferments are classified as : 1 Organized, as yeast.

2. Unorganized, such as occur in the animal body. They are characterized by the following : (a) small amount required for action ; (b) are not destroyed by the process ; (c) efficient only within narrow limits of temperature ; (d) they act only in a medium of a definite chemical reaction.

2. Coagulation is due to the action of a fibrin ferment on fibrinogen alone (Hammarsten).

Recent confirmation of the latter by the study of the blood of the tortoise (Howell).

Bearing upon the subject of coagulation of the investigations of Wooldridge.

The study of morphological elements of the blood other than the red and white corpuscles (Bizzozero, Osler, Hayem, Kemp).

Comparative: Investigation of the blood of crustacea, etc. (Halliburton, Howell).

It follows that fibrin does not exist preformed in living blood.

Appearances of the clot under different circumstances.

Variation in color; "the buffy coat."

Importance of coagulation in accidents; surgical operations, etc.

Paraglobin found in serum: how separated.

Fibrinogen obtained from transudation fluids.

Method of preparation.

Denis' *plasmine;* preparation.

Fibrin ferment; preparation.

Prevention of coagulation by cold. Settling of the corpuscles, (horse's blood).

Coagulation of the supernatant *plasma* on raising the temperature.

Coagulation by mixing serum and hydrocele fluid.

Coagulation in the heart and vessels of a dying animal.

Coagulation on *foreign* bodies placed in the vessels.
Why blood does not coagulate in the living body.
Fluidity of blood after death.

The influence of the living tissue of the internal coat of the blood vessels; effect of disease and injury of the same.

The influence of surface on coagulation.
The influence of the constant interchange in nutrition.
Circumstances favoring coagulation.
Circumstances retarding coagulation.

HISTORY OF THE BLOOD CORPUSCLES.

1. *Colorless* corpuscles originate probably in all leucocytenic tissues, especially in the lymphatic glands. Also, possibly, from division in the blood.

2. *Red* corpuscles; origin in the embryo.

Transition forms in red marrow.

From leucocytes.

In spleen and liver (?)

Fate of the corpuscles:

All have but a limited period of life.

They may, like the other constituents of the blood, be rapidly reproduced, when lost, as after hæmorrhage.

Both spleen and liver may be the sepulchres and transformation sites of the red corpuscles.

The pigments of the body are all, probably, largely derived from the coloring matter of the red corpuscles.

Dissolved corpuscles. "Laky" blood.

CHEMISTRY OF BLOOD.

Specific gravity of human blood, 1055.

Reaction, alkaline.

Corpuscles form about one-third of the entire blood.

About 90% of the plasma is water.

About the same proportion of the red corpuscles is *hæmoglobin*.

The most abundant salt of the blood is Na Cl.

Sodium salts are more abundant in *plasma*; *Potassium* salts in the *red corpuscles*.

Phosphates are also more abundant in the red corpuscles.

The proteids of the blood.

The coloring matter in vertebrates is embedded in the stroma of the corpuscles. Among invertebrates it is sometimes associated with the plasma.

In 100 parts of blood there are 12-15 parts of hæmoglobin. (*Gamgee*).

The large variety of substances found in the blood in small quantity.

The constantly varying composition of the blood.

PROPORTIONAL DISTRIBUTION OF THE BLOOD IN THE BODY.

About $\frac{1}{13}$ of the body-weight is blood, distributed as follows:

One-fourth in the lungs, heart, large arteries and veins.

One-fourth in the liver.

One-fourth in the skeletal muscles.
One-fourth in the remaining parts.
Abnormal: (clinical and pathological).
Transfusion: its advantages and dangers; plethora; anæmia; chlorosis; hæmorrhagic anæmia; leukæmia; scurvy; blood poisons, etc.
Inflammation; aneurisms; thrombosis; embolism.

DEMONSTRATIONS.

Blood issuing from an artery allowed to coagulate spontaneously in a receptacle.

After examination of the clot, serum, etc.

Defibrination of a portion of the blood in the above case.

After-examination of the washed fibrin.

Coagulated horse's blood, showing buffy coat.

Horse's blood drawn into a vessel surrounded by ice; non-coagulation, sinking of red corpuscles.

Clotting of plasma from the foregoing.

Vertical section of clot of horse's blood.

Coagulation by mixture of serum and hydrocele fluid.

The proteids of serum and plasma.

Dialysis of blood.

Coagulation under mercury.

Coagulation seen with the microscope.

Crystals of hæmin and hæmoglobin.

Heart clots; clots in the blood vessels.

Clotting on a foreign body introduced into a blood-vessel of a living animal.

Chemical examination of ashed blood.

LYMPH.

The tissues are not nourished directly by the blood but by lymph, which consists in great part of what has passed through the *smaller* blood-vessels by diffusion.

Lymph also contains matter derived from the tissues, the result of metabolism. The tissues at once select from and add to this fluid.

In the thoracic duct the lymph is mingled with the products of digestion or *chyle*.

Leucocytes abound in the lymph.

Lymph coagulates, giving rise to *fibrin*.

Circuit of the leucocytes in health.

Emigration of leucocytes in health; in inflammation.

Circumstances regulating the diffusion of lymph.

The rejuvenescence of lymph and blood.

Abnormal: Inflammatory and other *exudations*.

V. The Contractile Tissues.

These include amœboid cells; ciliate and flagellate cells; muscle cells.

Contractility is a property of protoplasm apart from the form it assumes.

The movements of amœba are *indefinite*, those of a muscle cell *definite*.

Both kinds of movements may result in *locomotion*:

All the movements of higher animals are due to the contraction of the protoplasm of muscles.

CILIATED CELLS.

Fixed, mostly columnar cells, with prolongations of the protoplasm from the free margin.

These *cilia* have a sickle-like motion, more rapid in one direction; give rise to currents in this direction.

The cilia are *automatic* in action; related to one another (coördinated); their motion extending from a definite point along the line of attachment of the cells.

Dependence of the movement on a certain temperature.

Influence of certain reagents.

Function and importance of ciliated cells in different parts of the human body.

The wide distribution and the importance of ciliated cells in lower animal forms.

Ciliated and flagellated animals and plants (cells).

Demonstration.

The rapid and energetic action of the cilia of the frog's alimentary tract (Bowditch.)

MUSCLES.

Muscle cells do *not* contract automatically; but they are *irritable*.

Divisions of muscles: 1. Plain (unstriated, visceral, organic, involuntary.)

2. Striated (skeletal, voluntary, striped). Among vertebrates plain muscle fibres are widely distributed, but the involuntary muscles in many cases are striated.

Histology: 1. *Non-striated* muscle cells: cells shorter, with well defined outline and elongated nucleus.

Arrangement into tissues.

2. *Striped muscle:* an entire muscle; bundles of fibres; connective tissue; tendinous attachments.

Composition of a single fibre.

Varying length and modes of attachment of the fibres.

A single muscle cell; sarcolemma; semifluid plasma; nucleus.

Alternating light and dim bands.

Krause's membrane.

Muscle as seen under polarized light; singly (isotropic) and doubly (anisotropic) refracting parts.

PHYSIOLOGY OF STRIPED MUSCLE; GENERAL.

The active and passive or contracted and relaxed condition of a muscle.

Variations in form and elasticity in each.

In the living body a muscle is, when relaxed or passive, under the influence of an obscure reflex (*tonus*).

The muscles are all in the passive state slightly *stretched*.

Importance of this to the animal.

Relaxation is the return to the passive condition, but is aided in most cases by the weight of the bony levers moved in contraction.

In contraction there is alteration of *form* without appreciable change in bulk.

A contraction is originated by a *stimulus*; passes in the form of a wave.

The changes in muscle are all associated with chemical activity.

The *physical properties* of muscles have been studied with special reference to *elasticity*.

Definition of elasticity.

The elasticity of muscle as compared with that of a steel spring. Curve of elasticity.

Muscle has slight but very perfect elasticity.

During contraction elasticity decreases, and extensibility increases. Diminishing lifting power. Advantages and disadvantages.

Elasticity conserves energy.

PHYSIOLOGICAL APPARATUS AND METHODS.

The study of muscle requires and illustrates the "graphic method" and the use of certain well known pieces of apparatus.

The graphic method as elaborated by Marey and other physiologists.

Its value and imperfections.

Photography.

Methods of registering time periods and movements.

Need of a standard of comparison.

Apparatus, &c.: Du Bois-Reymond's Inductorium.

Batteries used in physiological experimentation.

Keys: single and double.

Currents: constant, interrupted; induced.

Electrodes: ordinary; non-polarizable.

Myographs: drum; spring myograph (DuBois-Reymond); pendulum myograph (Fick).

PHYSIOLOGY OF STRIATED MUSCLE—(*Continued.*)

Muscular *irritability, contractility* and *conductivity*

A slight *stimulus* may give rise to a disproportionate effect (contraction).

Irritability is diminished by fatigue, imperfect blood supply, lowered temperature, and by certain drugs and poisons.

Evidence in favor of the *independent irritability* of muscle: contraction of muscle devoid of nerves (Sartorius); of embryonic muscle before nerves are developed; under *curare* poisoning.

Muscular *stimuli* classified as: mechanical, thermal, chemical, electrial, and, as in the body, nervous.

A single twitch or single muscular contraction follows a single stimulation.

Phases of a single contraction: latent period: phase of contraction; phase of relaxation. Average duration of each. $t = \frac{3}{7}, \frac{7}{5}, \frac{1}{\sigma\nu}, \frac{4}{10\nu}, \frac{1}{\nu}$

Circumstances causing these phases to vary.

Curve of a single contraction.

General law of contraction: A contraction ensues only on a *variation of intensity* of the stimulus. (0, +, —).

Variations according to the kind of tissue stimulated.

Latent period: its significance; period when changes electrical, chemical, thermal, and possibly mechanical, are taking place.

The contraction is the outcome of these changes.

Contracture (elastic after-action, contraction remainder).

Maximal, submaximal, minimal and subminimal stimuli.

Fatigue: associated with accumulation of waste products.

Effect of diminution in food supply (lymph).

Effect of venous blood; of saline solution.

Influence on the contraction curve.

Muscular work:

The work done is represented by the product of the weight into the height to which the latter is lifted (W = w × h.)

The lifting power of a muscle varies with the *area* of its transverse section.

The degree of shortening of a muscle varies with the *length* of its fibres.

The work done by a muscle increases with the load up to a certain point, then diminishes with increase of the load.

TETANIC CONTRACTION: PHYSIOLOGICAL TETANUS.

Tetanus: How produced.
  Perfect (smooth) and imperfect (vibratory) tetanus.
  Curve of tetanus.

Curve of *summation* of single contractions.

In tetanus the muscle may shorten to *one-third* of its original length.

Mechanical advantages of tetanus for the animal.

The normal muscular contractions of the body are tetanic and probably submaximal.

The muscle note.

Does its pitch correspond with the number of vibrations of the contracting muscle? No

Recent investigations on this topic (Yeo; Schäfer.)

Successive contraction of the fibres of a muscle in tetanus.

During the contraction of a muscle its blood vessels *dilate*.

Changes in Microscopic structure of a contracting muscle.

The *Isotropic* and *Anisotropic* substances remain distinct; but the anistropic seems to increase in bulk at the expense of the isotropic.

ELECTRICAL PHENOMENA OF MUSCLE.

1. *Resting Muscle:* Currents pass from equator to poles of a transverse section of a muscle.

Exist also in uninjured muscle in the body (Du Bois-Reymond).

Such do not exist in uninjured muscle (Hermann).

Evidence for each view.

All physiologists are agreed that there are electric currents in active muscle.

The galvanometer.

"Currents of action" (Hermann). "Negative variation" (Du Bois-Reymond).

The part of a muscle excited by a stimulus is electro-negative to the other parts.

The electric currents are associated with if not dependent upon *chemical* changes.

All the electrical phenomena are manifested during the *latent* period.

The "*Rheoscopic frog.*"

Secondary contraction set up in a "muscle-nerve preparation" (frog) by the beat of the mammalian ventricle.

LIVING AND DEAD MUSCLE COMPARED PHYSICALLY AND CHEMICALLY.

Living Muscle:

Soft, elastic, glistening, translucent and alkaline or neutral in reaction.

Active muscle is less elastic than passive muscle; more extensible; acid in reaction.

Dead Muscle:

Dull, opaque, inelastic; acid in reaction. Muscle gradually loses irritability when dying; *rigor mortis* (death) is sudden; is accompanied by the formation of CO_2 and sarcolactic acid.

Rigor Mortis causes shortening; relaxes when chemical decomposition takes place.

CHEMISTRY OF ACTIVE MUSCLE.

Excised living muscle contains no *free* oxygen.

Can contract in an atmosphere free from oxygen (H).

CO_2 and CH_3. $CH \begin{cases} OH \\ CO_2H \end{cases}$ (sarcolactic acid) formed.

In both the active and passive conditions muscle consumes O and forms CO_2; but these changes are more marked in the active state.

The weight of substances soluble in water decreases, and of those soluble in alcohol increases, *during action*.

Extractives of muscle.

Muscle at rest evolves heat; when contracting still more.

The steam engine and muscle compared.

LIVING MUSCLE CONSIDERED CHEMICALLY.

Muscle Plasma is semifluid.

Artificial preparation of muscle plasma.

Its coagulation; comparison with blood clotting.

Myosin corresponds to *fibrin*.

Myosin formed in *rigor mortis*; associated with the acid reaction.

Glycogen changed to sugar on the death of the muscle.

DEAD MUSCLE CONSIDERED CHEMICALLY.

Water 75 %.

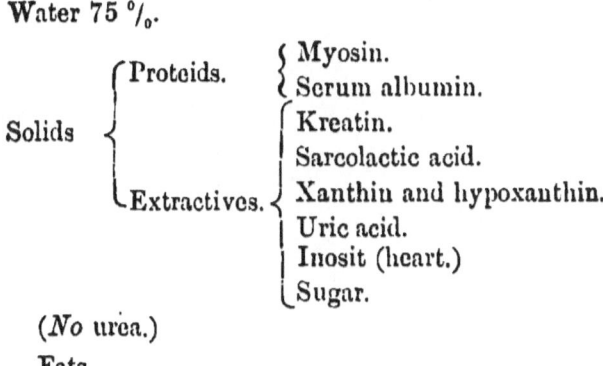

(*No* urea.)

Fats.

PHYSIOLOGY OF UNSTRIATED MUSCLE.

Apparent automatism of unstriated muscle. Where found.

The contraction of unstriated muscle is characterized by a long latent period and slowly-progressive wave.

The wave, unlike that of striated muscle, travels both longitudinally and tra versely along the cells.

The *galvanic* (constant) current is a more effective stimulus than the *induced* current.

Abnormal: comparison of the enlargement following systematic exercise with the diminution (atrophy) from disuse or disease.

The atrophy following separation from the nervous system.

Fatty degeneration.

VI. The Nervous Tissues.

These include the nervous *centres*, nerve *fibres* and modified nerve *endings*.

HISTOLOGY: nerve trunk: *funiculi*; *neurilemma*.

Nerve fibres: (*a*) medullated (white); (*b*) non-medullated (gray, pale or sympathetic; fibres of Remak.) Distribution of each kind.

Medullated fibre: Sheath of Schwann; medullary sheath; axis cylinder; nodes of Ranvier; nuclei.

Significance of the nodes.

The essential part of a nerve is the axis cylinder.

It is protoplasmic.

The other parts are protective, and the medullary sheath possibly nutritive.

These parts are lost before the nerve terminates peripherally or centrally (in many cases.)

Nerve-endings of muscle.

Relation of the sympathetic (pale) fibres to *unconscious* function.

Nerve centres: fibres; connective tissue (neuroglia); cells.

Forms and connections of the cells; *ganglia*.

Nerve endings: In most cases an axis cylinder associated with modified epithelial cells.

Various kinds: end-plates; Pacinian bodies; tactile corpuscles of Meissner; end bulbs of Krause; numerous forms in the organs of special sense.

Electrical fishes: gymnotus (eel), torpedo (ray).

Classification of nerves according to their functions (Martin).

Peripheral nerves.
- afferent
  - Sensory.
  - Reflex.
  - Excito-Motor.
  - Vaso-Motor.
  - Inhibitory.
- efferent
  - Motor.
  - Vaso-Motor.
  - Secretory.
  - Trophic?
  - Inhibitory.

Intercentral nerves.
- Exciting.
- Inhibitory.
- Commissural.

PHYSIOLOGY OF THE NERVOUS TISSUES.

Nerves have in a high degree, *irritability* and *conductivity*, but not automaticity.

The physiology of a nerve is largely a repetition of that of muscle, thus:

The *stimuli* of nerve are: mechanical, chemical, electrical.

Nerve is excited only when there is a *variation* in the stimulus.

Electricity is the most convenient form of stimulus for both muscle and nerve.

Nerve is more excitable to most stimuli than muscle.

ELECTRIC CURRENTS OF NERVE.

These follow the same laws in general as those of muscle; but while the wave passes over a contracting muscle at the rate of 3 metres in a second (frog), a nervous impulse travels at the rate of 28 metres in the frog, and 33 in man.

A "nerve impulse."

"Tetanization" of a nerve.

Certain drugs and poisons, change in temperature, etc., cause the *irritability* of a nerve to vary.

Nervous activity is probably attended by chemical changes and the evolution of heat, but such has not yet been demonstrated.

Electrical fishes.

DEMONSTRATIONS OF THE PROPERTIES OF MUSCLE AND NERVE.

Automatic contraction of the unstriated muscle of intestines of rabbit or frog.

Graphic representations of:

A single muscular contraction by direct and by nerve-stimulation.

A tetanus, perfect and imperfect. Contraction remainder. Secondary tetanus.

Curve of exhaustion.

Influence of temperature on a muscle contraction.

Maximal and submaximal contractions.

The contraction of a muscle with the use of the various kinds of stimuli.

The application of the constant current, and the induced current.

Elasticity of a steel spring compared with that of muscle.

Extensibility, without elasticity, of dead muscle.

Influence of the load on the work of the muscle.

Poisoning by curare.

Independent irritability of muscle (curare.)

The Rheoscopic frog.

Secondary contraction of the frog's muscle in response to the beat of the mammalian heart.

Illustration of degrees of stimuli: minimal, subminimal, etc.

Greater irritability of the "muscle-nerve preparation" than of curarized muscle.

Electrotonus.

Direct stimulation of a muscle attended by the same phenomena as stimulation of its nerve.

Comparison of muscular and nervous energy.

Greater vitality of muscle than of nerve.

Reaction of fresh, tetanized, and dead muscle.

Preparation of muscle plasma.

Preparation of myosin.

Preparation of syntonin.

Examination of watery extract of muscle.

ELECTROTONUS.

A muscle contracts with the passage of a *constant* current, only at make and break or other variation of the current; but the *irritability* of the nerve is modified Electrotonus).

How shown. Diagram.

Irritability is *increased* in the vicinity of the *kathode* or negative pole (*katelectrotonus*).

Irritability is *diminished* in the vicinity of the *anode* or positive pole (*anelectrotonus*).

Neutral or indifferent point.

The phenomena more marked the stronger the current.

Pflüger's diagram.

"Electrotonic currents."

"Law of contraction."

Bearing of the facts of this chapter on the treatment of certain diseases (electro-therapeutics).

Abnormal: Results following the section of nerves.

The Ritter-Valli law: centrifugal degeneration. Degenerative and regenerative changes of nerves.

THE FUNDAMENTAL PROPERTIES OF NERVOUS TISSUES.

Absence of nervous tissue in the *Protozoa*.

Beginnings in the cœlenterates (*Hydra*).

A nerve is a strand of modified protoplasm.

Development to a high degree in some divisions of the cœlenterates *(Medusæ).*

Progressive evolution of the nervous system leading to greater complexity and concentration.

Conductive nerve tissues; centres and centres.

Co-ordination leading to harmony in the whole life of the animal.

Diagramatic representations.

AUTOMATISM.

Automatic compared with reflex action.

The spinal cord.

Lymph hearts.

Peristalsis of the alimentary canal.

Peristalsis of the ureter (Engelmann).

The heart beat.

REFLEX ACTION.

General structure of the spinal cord; cells which are central, and fibres.

The spinal cord may be regarded as the seat of a great number of reflex centres.

The afferent and efferent fibres are mixed in a nerve trunk, but before joining the spinal cord, separate into afferent (posterior), and efferent (anterior) roots.

Anatomical requirements for reflex action:

1. An *afferent* nerve.
2. An *efferent* nerve.
3. A nerve *centre* which may be reduced to a single cell or to a sensory and a motor cell.

Diagramatic representation.

The afferent nerve terminates peripherally in "end-organs" (sense organs).

The study of reflexes is most conveniently accomplished in the frog.

Deductions therefrom:

The *latent* period of stimulation.

Purposive character.

Crossed or diagonal action.

Overflow, diffusion or radiation of the nervous impulses.

Lines of least resistance.

The reflex is the outcome of the action of the stimulus on the *central cells.*

A reflex may be co-ordinated or inco-ordinated.

The latter may be induced by the *direct* stimulation of the afferent nerve.

Importance of end-organs.

Each kind responds to its appropriate stimulus only.

The cord is not conscious.

Variations in the reflex depending on the kind and degree of stimulation.

Reflex time; greater part consumed by the cell processes.

Reflexes originate essentially independently of the will, but may be controlled or modified by conscious or unconscious cerebration.

· Inhibition of reflexes.

Reflexes have their expression in muscular action, secretion, etc., in which *efferent* nerves are *finally* concerned.

Influence of *strychnia* on the reflex action of the spinal cord.

Automatic and reflex action of central nerve cells.

The *sporadic ganglia* are probably not reflex centres.

Peculiar reflexes: Tendon reflex, cremasteric reflex, etc.

Muscular tone.

Protective value of the reflexes; reflexes in organic processes (secretion, etc., etc.)

INHIBITION.

The slowing or arrest of a process already begun is termed *inhibition*.

Inhibitory influences may proceed along an afferent nerve or from one part of a centre or collection of centres (brain; cord) and influence processes initiated in the same or other centres.

Examples of inhibitory action:

The *heart* by the pneumogastric nerve, and cardio-inhibitory centre.

The *respiratory* movements through the same nerve acting on the respiratory centre.

The inhibitory impulse may originate peripherally or centrally.

Inhibition artificially induced.

Importance of *inhibitory* impulses in *controlling* the activities of centres, and co-ordinating them for the general good of the economy.

DEMONSTRATIONS.

Comparison of a decapitated frog with one in which the brain and cord have been destroyed.

Comparison of the behavior of an intact frog with a decapitated one when kept in water which is gradually heated.

Reflexes of the decapitated frog:
Stimulation of an afferent nerve.
Immersion of the foot in acidulated water.
Application of bibulous paper moistened in acidulated water to different parts of the body, exciting reflexes of a purposive and co-ordinated character.
Radiation of nervous impulses.
Inhibition of reflexes.
Diagonal or crossed action.
Reflexes by various kinds of stimuli.
Reflexes with different degrees of stimulation.
Modification of reflexes by the administration of strychnia.
Functions of the roots of spinal nerves.

VII. The Circulation of the Blood.

ANATOMICAL AND PHYSIOLOGICAL.

1. *Comparative*: contractile *vacuoles* of the *protozoa;* contractile *tubes* (blood vessels) of *annelids;* the *reversing* heart of *ascidians;* heart (modified tube) of the scorpions; heart of *crustaceans;* heart of mollusks (oyster, clam, snail); circulation of *amphioxus;* heart of *fish;* heart of *amphibian* (frog); heart of *reptile* (tortoise, alligator); heart of *bird* and *mammal.*

The gradual rise in complexity of the vascular channels, etc., in the above forms.

Correlation with the blood, the respiratory organs, and the place in the scale of functional importance of the animal.

The best blood is always supplied to the head parts of the animal.

2. *The Mammal with special reference to man.*

Position of the heart in the unopened chest; supported by vessels and pericardium.

Pericardium and endocardium; pericardial fluid; importance of these.

Double character of the mammalian heart.

Relative thickness of the walls of its cavities; relation to function.

The skeleton of the heart.

Arrangement of muscular fibres.

Valves: semilunar; coronary openings; cusped valves: where strongest flap; *chordæ tendineæ*; *musculi papillares:* function.

Peculiar pocketing action of semilunar valves.

The muscular fibres of the heart blend with those of the great veins.

Pulsation of latter near the heart in the mammal imperfectly marked; very prominent in lower forms as in the *chelonians*.

Blood supply of the heart itself.

Nervous supply of the heart:

1. Vagus fibres some of which are efferent inhibitory, etc., others are afferent ("depressor.")

These are connected with the medulla oblongata.

2. Fibres from the spinal cord to sympathetic ganglia, and finally to the heart.

These are afferent and functionally "accelerators."

HISTOLOGICAL AND PHYSIOLOGICAL.

The peculiarities of the heart *muscle cells;* striated, nucleated, without sarcolemma, branched, etc. Intermediate character of function.

Distribution of elastic and fibrous tissue in heart and pericardium. Functional significance.

Nerve cells: In animals with a *sinus venosus*, it has a large supply of ganglion cells (Remak's); in mammals the two largest are near the orifice of the *superior vena cava*.

In cold-blooded animals as in the frog—other ganglia are found along the septum and on the auriculo-ventricular groove.

THE BLOOD VESSELS: ANATOMICAL AND PHYSIOLOGICAL.

Arteries, veins and *capillaries* compared. Arteries thick walled; lie open when cut; *veins* collapse, have valves.

Capillaries are microscopic.

The coats of a typical artery: best marked in a vessel of moderate size; elastic tissue most pronounced in the larger vessels; muscular tissue in the smaller.

Capillaries consist of a single layer of epithelium.

The marked elasticity of arteries; dilatibility of veins; power to vary in calibre of arterioles; fitness for the nutrition of the tissues of capillaries.

Importance of valves in veins.

THE PHYSICS OF THE CIRCULATION.

The object of the circulation of the blood is the *supply* of *nutriment* to all the tissues of the body as *required*, and the removal of waste matters.

The mechanisms concerned are the heart, arteries, veins and capillaries, under the control of the nervous system.

The nutriment of the tissues is derived finally from the lymph diffused through the delicate capillary walls.

The heart's action begins the circulation; the larger arteries store up the force expended by the heart and exert it between its contractions; the arterioles *regulate* the supply to different organs; the blood is delayed and diffused around the tissue elements while in the capillaries; gathered up and carried to the heart by the veins.

The vascular system may be compared to two trees united by their leafless branches. Comparison of two funnels.

The speed of the blood column is least when most divided as in the capillaries.

The sum of the united areas of the cross-sections of the capillaries is several hundred times that of the aorta.

Study of blood flow in frog's web, tongue, mesentery or lung: The "axial" current; the "inert" layer; movement of the red and the colorless corpuscles.

Abnormal: The circulation in inflammation as observed in the mesentery of frog (Cohnheim.)

Migration of leucocytes; diapedesis, &c., &c.

The circulation of the blood is effected by the following mechanisms:

1. A pump, intermittent in action.
2. A set of elastic tubes. Both acting on an incompressible fluid.

Comparison of flow from a severed artery and vein.

Influence of muscular exercise on the venous flow.

Blood-pressure.

What it is. How measured. Positive and negative. Mean blood-pressure.

Apparatus: Kymograph (Ludwig); Manometers mercurial; spring (Fick.)

Comparison of arterial, venous and capillary presure.

The *continuous* character of the vascular flow.

Importance of the elasticity of the arteries: "storage" of force.

Friction: how diminished.

Resistance chiefly *peripheral;* result: over full vessels; stretching.

The *veins* can contain all the blood of the body.

Large capacity of the *splanchnic area.*

Blood pressure is the *immediate* cause and the heart's action the *remote* cause of the vascular flow.

The velocity of the blood current.

The circuit of the circulation occupies in man about 23 seconds.

How determined.

Rule: The mean time required for the circulation is that occupied by 27 heart beats of the animal (Landois.)

Instruments for measuring the velocity of the flow: hæmadromometer (Volkmann); Stromuhr (Ludwig); hæmatochometer (Vierordt.)

THE PULSE.

The *pulse-wave* as distinguished from the blood current.

Its *velocity:* reaches the most distant part of the body before the ventricular systole is completed.

The characters of the pulse-wave vary with the nature of the vessel-wall.

Study of pulse tracings by an artificial *schema*.

Primary and secondary waves.

Tracings of pulse-waves.

Apparatus employed: sphygmographs.

Tracings: sphygmograms; hæmautograms.

The dicrotic, predicrotic and anacrotic notches.

The curves of high and low tension.

The pulse varies in rate and character according to: age, sex, position, time of day, phase of respiration, temperature, amount of exercise, emotional state, etc., etc.

The ratio of pulse rate and respiration in man (4 or 5 to 1.)

Abnormal: Variations of pulse with plethora, anæmia, hæmorrhage, fever, inflammation, &c.

The energy of motion lost to the blood in the circulation is restored in the form of heat serving for the maintenance of the body temperature.

THE PHYSIOLOGY OF THE HEART.

The cardiac cycle:

Study of the slow-beating heart of the frog.

The *peristaltic* nature of the contraction.

Cycle: systole, diastole, pause.

The passive stage (diastole and pause) occupy as much time as the systole of the whole heart.

Position of the mammalian heart in the body.

Position of human heart.

Movements of the heart in action: Its axes and changes in length; rotation. "So far as the ventricles are concerned the chief change during systole is one from a roughly hemispherical to a more conical form, effected without any marked diminution of the distance between the apex and the ventricular base" (Foster.)

The cardiac impulse. The hardening of the ventricles in systole; pressure against the chest wall.

Tracings of the events of a cardiac cycle by Marey and Chauveau.

The *cardiograph* (Burdon-Sanderson); Marey's *tambour*.

The systole of auricles and ventricles compared. Contraction begins in great veins. Persistence of the ventricular systole in complete contraction; more complete emptying of the ventricles than auricles.

The work done by the heart:

This depends upon: 1. The quantity of blood pumped out within a given time. 2. The resistance overcome.

Calculation of the work of the heart for 24 hours (124 foot-tons.)

INTRACARDIAC BLOOD PRESSURE.

How measured. The maximum manometer (Goltz and Gaule).

Influence on an empty pulsating heart (frog) of a neutral non-nutritive fluid (saline solution), showing that *intracardiac pressure is a stimulus to the heart.*

Within the body increase of intracardiac pressure does not always lead to increased *rhythm*. Such a case is complex.

Negative pressure within the auricles.

Effect on the working efficiency of the heart.

Negative pressure within the ventricles.

Diastole favored by the coronary circulation; economy of force.

Feeding the excised ventricle (frog, &c.) by the "perfusion canula" connected with various kinds of apparatus.

THE SOUNDS OF THE HEART.

Character of each.

Where best heard in man.

Theories as to causation of first sound:

1. Muscular origin (Ludwig & Dogiel).
2. Of wholly valvular origin.
3. Mixed origin.

Recent investigation of the cardiac sounds (Yeo).

The second sound is acknowledged to be entirely of valvular origin.

THE MECHANISM OF THE HEART BEAT.

Automatism of the heart.

Action of the isolated frog's heart.

The force of the heart beat is independent of the stimulus, *i. e.* minimal stimuli are maximal in effect.

The behaviour of the isolated apex.

The "stair-case" of beats (Bowditch).

Luciani groups. Diagram.

The pulsation of parts of the hearts of cold-blooded animals, independently of the rest of the organ.

Modifications of the heart beat:

These may be effected by: 1. The action of certain drugs and poisons operating through the blood or directly.

2. Changes of temperature.

3. Changes in *intracardiac* blood pressure.

" The rate of the beat is in inverse ratio to the arterial pressure " (Marey).

4. By nerve impulses originating either centrally or peripherally (reflexes).

Inhibition of the beat.

Stimulation of the peripheral end of the divided vagus causes slowing or total arrest of the heart's action in *diastole.*

Graphic tracing (frog's heart).

Condition of the heart during arrest.

Prevention of arrest by poisons (atropin).

Arrest by pilocarpin ; excitation by atropin.

Fibrillar action (Kronecker-Schmey phenomenon); negative influence of vagus over it.

Reflex inhibition: impulses act on the cardio-inhibitory centre situated in the medulla; proceed along various afferent nerves in most cases to the spinal cord primarily.

There is evidence that in certain animals, at least, the centres exercise a *constant* controlling influence as shown by the acceleration of pulse following section of the vagi nerves (dog).

Accelerator Nerves:

(*a*). Derived directly from the sympathetic ganglia: either the last cervical or last two cervical, and the first thoracic.

(*b*). Their origin may be traced to the spinal cord.

Stimulation of their peripheral ends causes *acceleration* and *augmentation* of the force of the heart beat.

When the vagus and the accelerator are stimulated simultaneously the action of the accelerator is masked; the heart is inhibited as by vagal stimulation alone.

The accelerator, as compared with the vagus, is marked by: the greater length of the latent period, the greater strength of stimulus required, and the greater persistency of action on cessation of the stimulus.

Accelerators are now more correctly denominated *augmentors*.

REGULATION OF BLOOD PRESSURE.

The factors concerned are:

1. *Work* of heart depending on force and frequency of beat, *i. e.* the quantity pumped into the arteries.

2. Peripheral *resistance* either of the smaller arteries (arterioles) or the capillaries. The greater the resistance the greater the blood pressure.

Accelerated action of the heart, as from stimulation of the *accelerators*, or by emotional states, does not raise the blood pressure, except momentarily.

Blood pressure is *lowered* by inhibition of the heart's action through stimulation of the pneumogastric nerve.

Tracing of vagal inhibition.

REGULATION OF BLOOD-PRESSURE BY VASO-MOTOR MECHANISM.

The *vaso-motor centre* is situated in the medulla oblongata.

"A small prismatic space in the forward prolongation of the lateral columns after they have given off their decussating pyramids" (Foster).

It includes the antero-lateral nucleus of Clarke, containing large multipolar cells; and is bilateral.

Secondary vaso-motor centres.

Blood pressure is not, except momentarily, lowered by moderate bleeding. Hæmorrhage sufficient to lower the blood-pressure sensibly is dangerous.

Injection of large quantities of fluid (blood, etc.), does not, except temporally, raise the blood-pressure.

Tonic contraction of *arterioles*.

How maintained.

Capillaries are probably under the control of the nervous system.

Section of the spinal cord causes extreme *lowering* of the blood pressure.

Stimulation of the peripheral end of the divided cord *raises* blood pressure greatly.

Regulation of local blood supply:

Determined by changes in the arterioles mainly.

Vaso-motor nerves:

Vaso-dilator: nervi erigentes; chorda tympani; nerves of muscles, etc.

Vaso-constrictor: splanchric; cervical sympathetic, etc.

Course of vaso-motor fibres.

Local apart from general vaso-motor tone.

Local vaso-motor paralysis by section of nerves.

Recovery of tone.

Influence of temperature; mechanical stimuli; drugs and poisons, on vaso-motor tone.

Impulses acting on the vaso-motor centre from higher centres. Blushing, pallor.

VASO-MOTOR REFLEXES.

Stimulation of the central end of a sensory nerve produces general rise in blood pressure owing to vascular constriction.

The *Depressor* nerve:

Runs separately from the vagus in some animals (rabbit, cat, etc.), but united with it in others.

When the vagi are cut and the central end of the divided depressor nerve stimulated there is rise of blood pressure independently of appreciably altered cardiac action.

This is owing to reflex dilation of the blood vessels of the *splanchnic area*.

Tracing of the above change in blood-pressure.

The reflex dilation of the rabbits ear from stimulation of the *auricularis magnus*.

Reflex dilation in the alimentary tract from the stimulus of food.

Rhythmic variations in the calibre of arteries and veins.

Vaso-motor effects on veins.

The co-ordination of the vascular factors for the general good of the body.

Abnormal: cardiac hypertrophy; atrophy; fatty degeneration of fevers. Disease of valves.

Disease of vessels: aneurism; atheroma; fibrosis; varicose veins.

Inflammation.

Recent investigations on the heart:

Several years ago it was shown that the *mammalian* heart can be isolated, and kept alive and in action for hours in favorable cases, by feeding with defibrinated blood and maintenance of a suitable temperature. (Martin).

Recent investigation of the depressor nerve of the mammal (Sewall).

The greater part of the most recent advances have been made by the study of the hearts of the cold-blooded animals (as yet) and have dealt with the *rhythm* and *innervation* chiefly.

Investigations on the rhythm and innervation of: the heart of the *frog* (Gaskell; Heidenhain); of the land *tortoise* (Gaskell); of the *terrapin* (Mills); of the *sea-turtle* (Kronecker and Mills; Mills); of the *fish* (McWilliam; Mills); of the *alligator* (Gaskell; Mills); of *Menobranchus* (Mills); of the *cuttle-fish* Ransom.

These investigations have greatly modified our conceptions of spontaneous rhythm; of the nature of the heart beat as a muscular act; of the nervous supply of the heart. Especially has it been shown that the *vagus* has functions much more varied and important than was formerly supposed.

Brief summary of the main conclusions.

DEMONSTRATIONS.

Flow from an artery and vein compared.

The action of the valves in the dead heart of a mammal.

The heart of the mammal *in situ*: chest opened.

Auscultation of sounds before and after opening chest; palpation of ventricles in action.

Isolated frog's heart:

The Stannius' ligature.

Arrest by electrical stimulation of sino-auricular junction.

Electrical stimulation of the isolated ventricle; "stair-case" of beats.

Pulsating sections of the heart of the frog.

Variations in the action of the heart with altered temperature; pilocarpin; atropin, &c., &c.

Behaviour of the ventricle of the frog's heart when supplied with nutriment.

Behaviour with altered pressure.

The work of the heart.

Inhibition of vagus effects by atropin.

Nerve effects:

Section of one and of both pneumogastric nerves of the dog.

Vagal inhibition in the mammal and frog.

Reflex inhibition in the frog (Goltz' experiment).

Vagus and accelerator effects on the heart of the mammal.

Stimulation of the accelerator.

Blood pressure:

Comparison of the arterial pressure in two large arteries of the same mammal.

Comparison of blood pressure in an artery and a vein.

Blood pressure experiments without the use of the kymograph.

Blood pressure experiments with use of kymograph; manometer and tracings.

Effects of:

Stimulation of peripheral end of vagus.

Stimulation of central end of depressor.

Section of spinal cord.

Stimulation of peripheral end of spinal cord.
Compression of the aorta.

Local blood pressure:

Effect on vessels of rabbit's ear of:
Section of the cervical sympathetic.
Stimulation of its peripheral end.
Stimulation of the central end of the *auricularis magnus.*
The sphygmograph: tracings.
Various points connected with blood-pressure, the heart, the pulse, etc., by simple apparatus improvised for the occasion.
Observation by the student in his own person of: pulse; cardiac impulse; study of the heart sounds with binaural stethoscope; venous flow, etc., etc.

THE LYMPHATIC CIRCULATION.

Anatomical and physiological:

Origin and structure of lymph capillaries: resemblances of lymphatics to veins.
Lacunæ or lymphatic spaces.
Principal points of junction of the lymphatics with the great veins.
Lymphatic glands: the regenerators of the leucocytes; the thoracic duct; main channel of communication with the digestive tract.
The lymph hearts of the frog, etc.
Lymph, chyle and blood compared.

Lymph is the *immediate* source of nourishment of the tissues which are bathed in it.

It is also the first recipient of the waste products of tissue metabolism.

The valves of lymphatics.

The onward flow of lymph is facilitated by:

1. Muscular exercise.
2. Aspirating influence of the great veins into which the main lymphatic trunks empty.

VIII. Respiration.

The purpose of respiration: removal of the waste-products of tissue metabolism; provision of oxygen for the tissues.

These ends are finally effected by *diffusion*.

The various mechanisms of respiration are subservient to the process of diffusion.

Evolution of Respiration: 1. Respiration by the general *surface* of the body as in *Protozoa*. 2. By the *integument* as in *Amphibia*.

3. Folded surfaces, external to the body, suitable for respiration in a liquid medium as by the *gills* of *Fishes*, and *branchiæ* of various invertebrates.

4. (*a*) Open tubes or *tracheæ* (insects). } Within the
   (*b*) Folded surfaces or *lungs*.   }   body;

Adapted for respiration in a *gaseous* medium as in *land vertebrates* and all *warm-blooded* animals.

The respiratory organs of the mammal are outgrowths from the digestive tract (fore-gut) of the embryo.

ANATOMICAL:

External respiratory openings (nares); trachea and bronchi; "bronchial tree"; folded membrane lined with flat epithelium (Pulmonary cells).

A *bronchus:* muscle ; cartilaginous rings (imperfect) ; mucous membrane with its mucous glands and *ciliated* epithelium.

The lungs: Subdivisions ; alveolar passages (infundibula); alveoli, air cells or air vesicles ; *flat* lining epithelium ; capillary circulation ; muscular and elastic tissue.

The relations of the *pleura* (closed sac) to lungs, chest walls, diaphragm.

The relation of lungs to chest walls, heart and great vessels diaphragm.

Comparative: The lung in the amphibian, reptile, bird ; air bladder of fishes, etc.

THE MECHANICS OF RESPIRATION: RESPIRATORY MOVEMENTS.

Marked *elasticity* of the lungs ; made up in great part of elastic tissue.

Never fully distended in the thorax.

The elasticity of the human lungs may be represented by a manometric column of 5 millimetres of mercury.

How measured.

Degree of pulmonary distention in ordinary inspiration, forced inspiration, ordinary expiration, forced expiration, and the death position compared.

Friction, how lessened.

In all the movements of the chest walls and diaphragm the lungs follow and are closely applied to their surfaces.

Why the lungs expand in inspiration (atmospheric pressure) and diminish in expiration (elasticity).

Inspiration is the *active* phase of respiration; expiration the *passive* one.

Recoil of chest walls.

Modifications of the shape and size of the chest during respiration.

The diameters of the thorax: vertical, transverse, antero-posterior.

Muscular mechanisms by which they are made to vary; muscular mechanics:

Muscles of fixation; of elevation (and depression); of eversion; ordinary and extraordinary.

Co-ordination of muscular action in an ordinary inspiratory act.

Physiological (functional) division of muscles according to the character of the respiratory act.

The scaleni, the intercostals, the diaphragm.

The diaphragm is the most important respiratory muscle.

Importance of the *abdominal* muscles in labored respiration, and various special modifications of the respiratory act associated with other functions (defæcation parturition, vomiting, sneezing, coughing, etc.)

The costal movements.

The lower ribs in forced inspiration.

Types of respiration in the human subject:

Thoracic, abdominal, (costo-superior, costo-inferior)

Thoracic and abdominal or diaphragmatic respiration, in other vertebrates.

Facial and laryngeal movements in man and other animals.

Friction, how lessened.

Respiratory rythm.

Inspiration to expiration as 10 to 12.

Curve of thoracic respiratory movements.

(Marey's Pneumatograph).

" Cardiac respiration " of hybernating animals.

Modified respiratory acts :

Sneezing, coughing, laughing, crying, hiccough, etc.

THE VOLUME OF AIR; ITS VARIATIONS.

A certain portion of the air of the lungs is moved by diffusion only. The moving column of air in ordinary respiration is of small volume.

Tabulation of volumes :

Stationary air { residual		100 cubic inches.
200 cubic inches { supplemental		100 cubic inches.
Tidal air.		30 cubic inches.
Complemental air		98 cubic inches.
Total capacity of lungs.		328 cubic inches.

Vital capacity; modifying causes.

" Post mortem " air; " tissue " air of collapsed lung.

The quantity of air requisite for daily consumption.

Method of computation.

CHANGES OF AIR IN RESPIRATION.

1. Difference in temperature of the expired and inspired air.

Heat is thus lost to the body to the extent of 20 per cent.: (a) by warming the air (5 per cent); (b) by evaporating the water given off with the expired air (15 per cent).

Variations in the above.

2. Difference in quantity of moisture. Expired air is almost *saturated*.

3. Difference in chemical composition.

The atmosphere is composed of oxygen "dissolved in" nitrogen.

Analysis of the air of respiration:

	Oxygen.	Nitrogen.	Carbon Dioxide.
Air of inspiration	20.810	79.150	.040
Air of expiration	16.033	79.557	4.380

Total quantity of O consumed, and CO_2 exhaled, in 24 hours.

Respiratory relations of animals and plants.

Proportion in volume of the inspired and expired air.

Real loss in volume by inspiration.

"Respiratory quotient" $\dfrac{CO_2}{O}\left(\dfrac{4.380}{4.782}\right)=0.906.$

Effete organic matter and volatile organic bodies of the expired air.

Ventilation. CO_2 standard of impurity.

Classification of gases in relation to the Respiratory function:

1. Indifferent: H, N.
2. Irrespirable (spasm); Cl, NH_3, &c.
3. Poisonous:
   (*a*) Narcotic CO_2, N_2O, Ozone.
   (*b*) Reducing H_2S, PH_3, AsH_3, SbH_3, CN, CO, NO.

THE CHEMISTRY OF RESPIRATION.

External (pulmonary) and internal (tissue) respiration:

Difference in color between arterial and venous blood.

Ordinary venous blood and the blood of asphyxia compared.

Average composition of the gases derived from blood:

	O	CO_2	N
Arterial blood.........	2 0	40	1–2
Venous blood.........	8–12	46	1–2

From 100 volumes of blood 60 volumes of gas may be obtained.

Apparatus for the investigation of the blood gases.

The laws governing the absorption of gases by ordinary liquids compared with those regulating the corresponding absorption by blood.

Law of *partial pressure.*

Interchange of gases according to the *tension.*

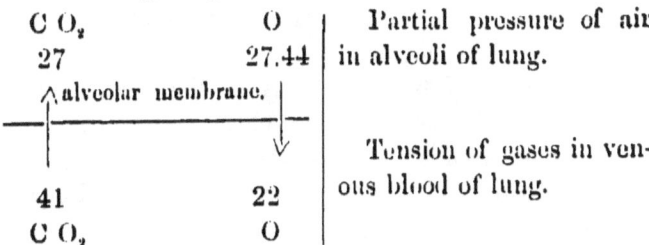

The oxygen of blood is almost all combined with the coloring matter of the red corpuscles, hæmoglobin.

The CO_2 is chiefly in loose chemical combination with certain constituents of the plasma.

The nature of the union between the oxygen and oxyhæmoglobin is one of "loose chemical combination."

Chemical constitution of hæmoglobin ; C, H, N, Fe, S, O, in complex combination.

Intra-molecular O ; loosely combined O.

There is a considerable quantity of iron in the molecule (0.4 per cent).

This Fe is replaced by other metals in some animals ; in which also the *plasma* and not the corpuscles seems to function as the respiratory part of the blood (Crustaceans).

In some animals Fe is associated with the plasma (annelids).

Hæmoglobin may be decomposed into a globulin and *hæmatin;* the latter contains Fe but no S.

Hæmatoin (Preyer) is hæmatin free from Fe.

Hæmin is Hydrochlorate of hæmatin.

Crystals of hæmin (Teichmann's crystals).

Crystals of hæmoglobin.

Importance of blood crystals in the recognition of altered blood; distinguishing blood and solutions of hæmoglobin from other red fluids, etc.

Hæmatoidin (bilirubin): crystals of old blood clots. Composed of C, H, N, O.

O—Hæmoglobin.

C O—Hæmoglobin.

N O—Hæmoglobin.

Poisoning by N O, C O, C N-hæmoglobin.

Arterial blood is normally almost or quite saturated with oxygen.

Venous blood always (except in fatal asphyxia) contains *some* oxyhæmoglobin.

The oxidative processes of the body occur in the tissues and not in the blood.

Illustrated by the study of muscle.

During the day time and during exercise, more oxygen is given off as CO_2, than is absorbed. The reverse holds during the night and during rest.

The same laws hold for the interchange of gases between the blood and the tissues as between the blood and the pulmonary air.

Relation of blood pigments to the other pigments of the body.

SPECTRUM ANALYSIS OF BLOOD PIGMENTS.

Spectrum appearances of a solution of:
1. Oxyhæmoglobin of different degrees of strength. Disappearance of one band.
2. Hæmoglobin (reduced hæmoglobin).

How distinguished from a concentrated solution of O—hæmoglobin.

3. C O—hæmoglobin (carbonic oxide hæmoglobin).

Its permanancy and irreducibility : will resist putrefaction (Hoppe-Seyler; Landois).

4. Acid and alkaline hæmatin.
5. Hæmachromogen (reduced hæmatin).
6. Hæmatoporphyrin.
7. Methæmoglobin (spectrum like that of acid hæmatin).

Methæmoglobin contains more oxygen than oxyhæmoglobin (Hoppe-Seyler).

Oxyhæmoglobin changed to methæmoglobin when discharged into urine, etc.

Recent spectroscopic investigations (MacMunn).

INFLUENCE OF RESPIRATION ON THE CIRCULATION.

Aspirating effect of the inspiratory movements.
Retardation of venous flow during expiration.

The negative effect on the great arterial trunks of the thorax.

Comparison of the curve of blood pressure with the curve of intra-thoracic pressure.

Graphic representation of changes in the pulse-wave. Traube-Hering curves.

These seem to be owing to rhythmic variations in the action of the vaso-motor centre.

The respiratory variations of blood-pressure as a whole must be referred to mechanical rather than to nervous causes.

THE NERVOUS MECHANISM OF RESPIRATION.

Respiration is involuntary; it may be *modified* but not wholly arrested by the will.

Anatomically the mechanism concerned in respiration consists as in other reflex acts of:

(*a*) *Afferent* nerves, especially the vagus.

(*b*) *Efferent* nerves, supplying the various respiratory muscles.

(*c*) A respiratory *centre* in the medulla.

There are probably other subordinate centres in the spinal cord. The phrenic arising from the cervical spinal cord is the great motor nerve of the diaphragm.

The respiratory centre or *nœud vital* (Flourens) lies below the vaso-motor centre and above the *calamus scriptorius*.

Experimental:

(*a*) Complete cross section of the spinal cord in different regions.

(*b*) Division of a costal nerve: paralysis of its muscle.

(*c*) Division of one and of both phrenic nerves; paralysis of diaphragm.

(*d*) Division of one and of both vagi: modified respiratory movements; deeper and slower respiration; marked pause.

Stimulation of: (*a*) central end of divided vagus; the other nerve being also cut: inspiratory tetanus.

(*b*) Central end of superior laryngeal: expiratory tetanus.

Division of the medulla in the middle line; *geminal* character of centre.

Increased action of the respiratory centre:

(*a*) After division of the spinal cord below the medulla oblongata with section of both vagi.

(*b*) After ligature of the blood vessels of the neck.

The increased action is due chiefly to diminution of the supply of oxygen to the centre.

The action of drugs on the centre through the blood.

Effect of emotions; of cold (water) suddenly applied to the skin.

Summary of inferences from the above experiments:

Impulses which tend to modify the discharges from the respiratory centre are constantly proceeding to it along the spinal cord, chiefly; but also from the cerebrum; either originating primarily in these centres or proceeding to them by various nerves, especially the pneumogastric.

The costal and the phrenic are the most important motor nerves of ordinary inspiration.

The pneumogastric is the path of impulses tending to quicken the respiratory movements and the superior laryngeal that of those of an opposite character.

The *respiratory centre* is geminal. It is essentially automatic in action; but its action is constantly being modified by afferent impulses.

It is possible that the movements of the lungs themselves may furnish stimuli exciting to action of the centre.

The action of the centre is determined by the quality of the blood supplying it; venous blood being stimulating (CO_2).

The nature of *dyspnœa apnœa* and *eupnœa* compared with special reference to the respiratory centre.

The *rhythmic* action of the respiratory centre is probably the resultant of two forces, one resisting, the other exciting to nervous discharges.

The action of the centre may furnish its own stimuli in the form of waste-products.

The nerves are at once the conductors of modifying impulses, and the avenues along which the discharges of the centre travel to the muscles.

The *resistance* theory of Rosenthal.

The double character of the centre: inspiratory and expiratory.

The phenomena of **Asphyxia**; stages and duration; character of the blood.

Post mortem appearances of the heart and blood-vessels.

Cheyne-Stokes respiration.
Respiratory sounds.
Effect on the respiration of changes of pressure of the air breathed.
1. Gradual diminution of pressure: dyspnœa owing to diminution of O.
2. Sudden diminution of pressure: liberation of gas, causing mechanical interference with the circulation.
3. Increase of pressure: when of a certain degree (4 atmospheres of O) leading to diminished oxidation with symptoms of asphyxia.
Artificial respiration.
Respiration of hybernating animals.

ABNORMAL:

Dyspnœa of asthma; feeble respiration of phthisis; hampering effect of pleuritic adhesions.
Pneumothorax; pleuritic effusions.
Effects of altered pressure: ascent in balloons; scaling mountains; descending shafts of mines.
Asphyxia: drowning, hanging, choking; poisoning by carbonic oxide.

DEMONSTRATIONS:

Anatomical examination of *gills* of oyster, crab, fish and menobranchus; respiratory *sacs* of menobranchus; *air bladder* of fish; *lungs* of frog, snake, bird, mammal.
Demonstration of the gaseous interchange in respiration.

Respiratory *schema*.

Measurement by the mercurial manometer of the relative pressure in inspiration and expiration, ordinary and forced.

The respiratory movements of the frog.

Phenomena of hyperpnœa, dyspnœa, apnœa, asphyxia, in a mammal.

Post mortem on an asphyxiated mammal.

Examination of spectrum of blood of the same.

Examination of the heart of the above before and after rigor mortis.

Comparison with the heart of an animal bled to death.

Facial and laryngeal respiration of the rabbit.

Section of the spinal cord of a frog.

Section of the various respiratory nerves of a mammal.

Stimulation of respiratory nerves of a mammal.

Puncture of respiratory centre.

Respiratory sounds.

Puncture of the thorax of a respiring mammal.

Artificial respiration.

Spectra of blood as indicated on page 60.

Tracing of the variations in the column of air in a respiring mammal.

Demonstration on the human subject of many points.

Self-observation by the student.

IX. Digestion.

THE GENERAL PHYSIOLOGY OF SECRETION.

Secretion is the resultant of forces, some of which are purely physical and others peculiar to living cells, the action of which is not yet explicable.

Laws of the diffusion of gases and liquids, as witnessed in the lungs; interchange between the lymph and the blood, etc.

The relations in secretion may be thus represented:

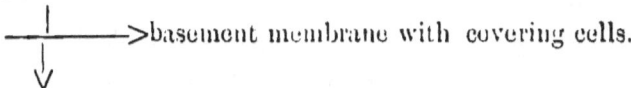

Secretion

The simplest gland is *unicellular*.

The most complicated gland is but an aggregation of cells, with special arrangements for holding them together, for blood supply, and for removal of their secretion.

The simplest form of gland in the mammal is the *tubular*; but there are *secreting membranes*.

Extension of surface provided by increasing complexity of structure.

In digestion the secretions act on food.

Characteristics of a *food*.

Food of animals and plants compared.

The food of animals must be organic together with water and salts.

The greater part is not only organic but *organized*.

The purpose of a food: to reconstruct tissues and restore energy.

Energy is derived chiefly from oxidizable compounds of carbon and hydrogen with or without nitrogen.

Certain substances important for the welfare of the organization seem to be chiefly *regulators* of nutritive processes.

"Adjuvants": relishes, condiments, stimulants.

CLASSIFICATION OF FOOD-STUFFS (INGESTA.)

Inorganic { Water. Mineral salts. { Soluble. Insoluble. } Incombustible.

Organic. { Nitrogenous. Non-nitrogenous. } { Albumins. Albuminoids. Carbohydrates. Fat. } { Sugar, Starches, etc. } Combustible.

Egesta represent altered food-products, useless or harmful to the economy.

The principal part of nitrogenous foods reappears in the *excreta* as crystalline nitrogenous bodies; that of the hydrocarbons and carbohydrates as carbon dioxide and water.

Importance of cooking.

Typical food: milk (and eggs).

DEMONSTRATIONS.

The principal properties and reactions of the above food-stuffs.

DIGESTION IN THE MOUTH.

Anatomical and physiological:
The *teeth* of different groups of animals compared.
Relation of structure to function.
The *tongue;* its function in different classes of animals.
Tentacles in the invertebrates.
Minute anatomy of:
Teeth: enamel; dentine; crusta petrosa; nerve and vascular supply.
Tongue: arrangement of muscles; nerve-endings.
Mucous glands: abundant distribution in the buccal and pharyngeal cavities.
Food in the mouth is comminuted and submitted to the action of saliva.
Mastication: muscles concerned.

SALIVARY GLANDS.

Mucous and serous glands: resemblance in minute structure to the Pancreas.
Belong to the class *racemose.*
Crescents of Gianuzzi; composed of *marginal* cells.

SALIVA.

Mixed saliva found in the mouth.
Secretion of *serous* and *mucous* glands compared.
Morphological elements of saliva.
Chemical constitution.
Bodies *peculiar* to saliva: (*a*) *Ptyalin*, nitrogenous but not a proteid; (*b*) *Potassium sulphocyanide*.
Ptyalin is an unorganized ferment (see page 27).

Digestive action of saliva:

Requires an alkaline medium.
Influence of temperature.
Ptyalin bears a somewhat high temperature, but not that of boiling water.
Rapid action of ptyalin.
The sole digestive action of saliva is on the starchy elements of food (diastatic action).
Starch converted into sugar: maltose and dextrose, chiefly the former.
Dextrin and acroodextrine.

NERVOUS MECHANISM OF SALIVARY SECRETION.

The nerve supply of the submaxillary gland: sympathetic and cerebral.

Experimental facts:

Stimulation of the peripheral end of the *chorda tympani* supplying the submaxillary gland) of the dog is followed by: (*a*) Dilation of the blood vessels of gland; there may be a venous pulse.

(*b*) Copious flow of saliva.

(*c*) This secretion will take place in a decapitated animal.

(*d*) Secretion follows stimulation of the lingual nerve or of the tongue (reflex act).

(*e*) After administration of atropin secretion does not take place, but the vessels dilate as before.

(*f*) Secretion will take place when the pressure in the duct is greater than that of the blood in the artery of the gland or even the carotid.

Conclusions:

1. The Chorda Tympani is the secretory nerve of the gland; secretion is *accompanied* by vascular dilation (*a* and *b*).

2. Secretion is not purely a process of filtration (*c*, *e* and *f*).

3. Secretion may be excited reflexly (*d*).

Stimulation of the cervical *sympathetic* in the dog gives rise to vascular constriction and scanty flow of viscid saliva; the exact opposite of the results on stimulating the *chorda*.

Diagram of the nervous mechanism of salivary secretion.

The "paralytic" secretion of saliva.

The secreting *centre*. Influence of higher centres over it.

The Gland compared histologically before and after stimulation:

Behaviour towards reagents; alteration in lumen; alteration in size of cells; position of nucleus; degree of granulation, etc.

The nervous mechanism of salivary secretion was worked out by Claude Bernard.

Constructive (*anabolic*) and destructive (*katabolic*) action in the gland protoplasm; *mother-ferment, pro-ferment, zymogen.*

DEGLUTITION.

The *bolus* of food; part taken by tongue, cheeks, &c.

Swallowing may be partly voluntary and partly involuntary; is essentially involuntary.

Stages; muscular mechanism.

Protection of the respiratory openings.

Is a reflex act; *centre* in the medulla.

The *œsophagus:* circular and longitudinal muscular fibres; voluntary and involuntary muscle cells; mucous glands.

Peristaltic action.

Relaxation of cardiac *sphincter*.

Swallowing as witnessed in the *ruminants*.

Force exerted by the gullet in swallowing (Mosso).

DIGESTION IN THE STOMACH.

ANATOMICAL AND PHYSIOLOGICAL:

The stomach in the frog; carnivore; ruminant; etc., compared.

The human stomach: form, relations, nervous and vascular supply.

Fundus; cardiac and pyloric ends.

Walls of empty stomach (and gullet) collapse.

Folds of mucous membrane (*rugæ*) of the collapsed stomach.

Coats of the stomach.

Arrangement of the muscular fibres.

Sphincters: Cardiac and pyloric (valve).

Nerves: vagus; splanchnic.

Nerve cells.

Gastric glands: kinds; differences.

The cells of the glands, (*a*) *central* cells *chief* cells, secrete pepsin and rennet.

(*b*) *Superadded, parietal* or *oxyntic* cells.

Glands of the pyloric end have only central cells.

Transition glands of the *pylorus*.

Arteries, after distribution in the *areolar* coat, break up as capillaries around the bases of the glands.

THE GASTRIC SECRETION.

How obtained. Fistulæ. Flow only during digestion.
Physical and chemical properties of the secretion.
The ferment is *pepsin*.

The acid reaction is owing to H Cl—present to the extent of 0.2 per cent.

The H Cl is free.

The presence of butyric, lactic and other acids is traceable to fermentation.

Quantity of gastric juice; variations during digestion.

Digestive action.

The activity of gastric juice is exerted only on proteids.

Action owing to *pepsin*: efficient only in an acid medium.

Auxiliary to pancreatic digestion.

The mucus of the stomach seems to contain a ferment capable of converting cane-sugar into dextrose.

$$C_{12} H_{22} O_{11} + H_2 O = 2 (C_6 H_{12} O_6)$$

Solution of salts in the H Cl of the secretion.

Pepsinogen: the *zymogen* of gastric juice.

Functions of the different kinds of cells of the gastric glands.

Rennet: acts on casein as a curdling ferment.

Has a curdling power independently of the acid of the stomach.

The conversion of proteids into *peptones* which are soluble and diffusible.

Other properties of peptones: soluble in water; not precipitated by boiling; give the reaction of proteids, etc.

Formation of *parapeptone* (acid albumin?)

Peptone may be formed by *hydration*.
Chyme: its crude character.
Antiseptic properties of gastric juice.
Artificial digestion.
Preparation of pepsin extracts:
    (*a*) Watery extract.   (*b*) Extract with dilute hydrochloric acid (O. 2 p. c).
    (*c*) Glycerine extract.   (*d*) Glycerine extract after dehydration by alcohol.

THE ACT OF SECRETION AND THE MOVEMENTS OF THE STOMACH.

During digestion the mucous membrane is flushed with blood.

Movements of the stomach cause currents in the food; their importance.

Inasmuch as normal secretion is poured out after division of all the nerves (vagus and splanchnic) the mechanism of secretion seems to be local, though it is probably under the control of the nervous system.

As digestion is affected by emotions, etc., there must be a nervous mechanism by which impulses are conveyed to the stomach.

The effects of section and stimulation of nerves.

Circumstances affecting gastric digestion.

The study of Alexis St. Martin (Beaumont).

Vomiting:

Pyrosis: simple character.

Vomiting a *reflex* act; centre in the *medulla*.

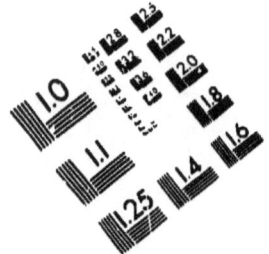 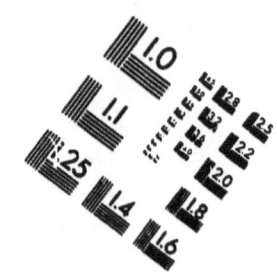

IMAGE EVALUATION
TEST TARGET (MT-3)

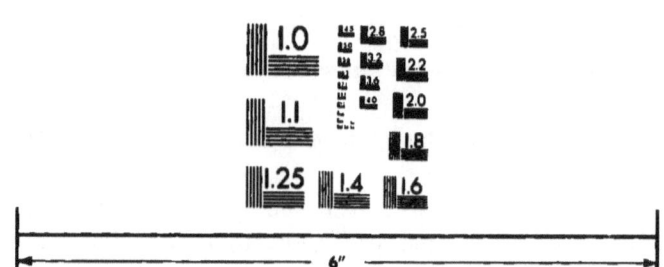

Photographic
Sciences
Corporation

23 WEST MAIN STREET
WEBSTER, N.Y. 14580
(716) 872-4503

Muscular mechanism by which the gastric contents are ejected.

The relaxation of the sphincters.

The part taken by the respiratory system.

Vomiting owing to direct action of the centre as by drugs in the blood.

Action of the higher centres on the vomiting centre (emotions).

Natural regurgitation of food (birds).

DIGESTION IN THE INTESTINE.

The arrangements in the intestine provide for a continuation of digestive processes already begun; the initiation of new ones; the absorption of fully digested material; and the ejection of what is undigested and useless.

The digestive secretions poured into the intestine are all alkaline and include: *pancreatic juice*; *succus entericus* and *bile*.

Anatomical and physiological:

The intestine in the carnivore and the ruminant.

Divisions of the intestine.

The coats: arrangement of the longitudinal coat in the large intestine into bands.

Valvulæ conniventes.

Nerve plexuses: Auerbach's between the two layers of the muscular coat.

Meissner's in the submucous coat.

Glands: crypts of Lieberkühn, in the whole of small intestine; are tubular.

Brünner's glands of duodenum are racemose; agminated or solitary glands; Peyer's patches, abundant in Ileum.

The two latter are composed mainly of lymphoid tissue.

Absorbent *villi*, over the whole surface of small intestine.

Structure of a villus: Columnar epithelium with *striated* border; basement membrane; capillary network; adenoid tissue; plain muscular fibres surrounding a central lacteal.

Numerous lymphoid cells in the adenoid tissue and at the bases of and between the columnar cells.

PANCREATIC DIGESTION.

Anatomical: The pancreas closely resembles the salivary glands in structure; *alveoli* longer and more tubular instead of being saccular (Schäfer).

The duct of the gland opens into the duodenum.

Pancreatic juice:

Physical and chemical constitution; alkaline reaction:

Solids (8 p. c.)
- Albumin.
- Alkali albumin.
- Peptone.
- Leucin.
- Tyrosin.
- Fats and soaps.
- Considerable sodium carbonate.

Peptone, leucin, tyrosin and soaps are due to digestive changes after secretion.

The large quantity of proteids is noteworthy.

Pancreatic juice is the most complex and important of all the digestive secretions.

It contains three ferments: *Trypsin*, acting on proteids; *amylopsin*, acting on starches; *steapsin*, acting on fats.

The *amylolytic* action is akin to that of saliva.

Neutral *fats* are partly *emulsified* and partly decomposed into free *fatty acids* and glycerine.

Proteids are converted into *peptones*.

Pancreatic proteid digestion differs from gastric in that it takes place in an alkaline medium; the final product is not peptone alone but also leucin and tyrosin; alkali albumin is formed instead of acid albumin; the formation of Indol.

Leucin and tyrosin are crystalline nitrogenous bodies (C, H, N, O).

Indol is not an essential product of pancreatic digestion.

Indol (C, H, N) appears in the urine in an oxidized form as Indican (C, H, N, S, O).

To Indol is due the fœcal smell of the intestinal contents.

Putrefaction of pancreatic digestion is owing to bacteria (formation of Indol, &c.)

Extracts of the pancreas when living; when dead, etc. etc., compared.

Pancreatic juice, like saliva, acts in a neutral but not in an acid medium.

Acts best in an alkaline fluid (sodium carbonate) of 1-2 per cent strength.

BILIARY DIGESTION.

The secretion of bile is constant; but it is stored up in the gall bladder during the intervals of digestion.

Anatomical:

The liver: How the cells are collected together into masses (*lobules*). How the blood supply is effected, and the secretion drained off and conveyed into the alimentary canal.

Technical: *Hilus; Glisson's capsule; portal* vein, *hepatic artery* and *hepatic duct* running together; the *interlobular, intralobular* and *sublobular* veins.

The *capillary plexus* around the cells within the lobules.

The intercellular bile *canaliculi.*

The *ductus communis choledochus.*

The *portal system* of vessels.

CHEMICAL AND PHYSICAL CHARACTERS OF BILE.

Reaction: alkaline.

Color: green in herbivora; yellow in carnivora; intermediate in man.

Pigments: Biliverdin, bilirubin.

Cholesterin: a fatlike crystalline substance, especially abundant in gall stones.

Bile salts: Taurocholate and glycocholate of sodium.

These acids are composed of cholalic or cholic acid, (non-nitrogenous) united with an amido-compound (NH_2).

The bile acids are composed of C, H, N, O and in the case of taurocholic acid S in addition.

Mucus derived from the gall bladder chiefly.

The absence of albumin is noteworthy.

THE SECRETION AND THE DIGESTIVE FUNCTIONS OF BILE.

Bile is secreted under a pressure *less* than that of the blood.

Reflex action of the gall bladder under the influence of the acid contents of the stomach when they meet the orifice of the bile duct in the duodenum.

Diagram of the rate of pancreatic and biliary secretion.

Precipitation of pepsin, peptones, parapeptones and bile salts, when bile or a solution of bile salts is added to the products of gastric digestion; excess dissolves.

Advantages: removal of pepsin, which is antagonistic to trypsin; prevention of the too rapid passage of the digestive mass along the intestine.

Bile possesses slight emulsifying powers; is antiseptic; a natural purgative; facilitates the absorption of **fats**; forms soaps which favor emulsification; is solvent of fats.

The action of bile is owing to the presence of the *bile acids*.

Biliary fistulæ.

Clay colored stools of jaundice.

Chyle, compared with *chyme*.

Recent investigation: (Landwehr). Bile decomposes mucus into animal gum and a globulin substance; animal gum forms an emulsion with fats readily.

Succus entericus: An alkaline fluid; has feebly amylolytic and proteolytic power.

Intestinal fistulæ.

Movements of the intestine.

Normal movements in the body.

End at Ileo-cœcal valve.

Movements of a piece of excised intestine.

Increased action under obstruction.

Increased action during asphyxia.

Impulses pass along the *vagi* and *splanchnic* nerves.

Central connection (emotions).

THE LARGE INTESTINE.

Anatomical and physiological:

The ascending, transverse and descending colon. Sigmoid flexure.

Tubular glands but no villi.

Little digestive action in carnivorous animals and man in the large intestine.

Cellulose digestion of ruminants.

Movements of large intestine begin at the Ileo-cœcal valve.

Fermentations, giving rise, together with pancreatic putrefaction, to the gases of the intestine : CO_2, CH_4, N, H, H_2S, etc.

These are found also in the small intestine.

Defæcation: comparison with vomiting, parturition, etc.

Muscular mechanisms.

The defæcating *centre* is situated in the lumbar cord in the dog (Goltz).

The essentially *reflex* character of the act.

Fæces:
- Undigested remnants of the *ingesta*.
- Ferments.
- Mucus.
- Bile acids.
- Altered bile pigments. Bilirubin or Hydro-[bilirubin, a "reduction" product.
- Cholesterin.
- Fatty acids.
- Insoluble soaps of Ca and Mg.
- Excretin (C, H, O, S) crystalline.

ABSORPTION.

Chyle: comparison with lymph and blood.

Fats: absorbed mostly through the villi and conveyed by the lacteals into the venous blood.

A small quantity by the portal circulation.

Recent investigation on fat absorption (Schäfer).

Proteids: mostly by the smaller blood vessels of the portal system.

In part by the villi (?)

Sugar: chiefly by the blood vessels of the portal system.

Absorption according to the laws of diffusion.
Osmosis: endosmosis; exosmosis; crystalloids and colloids.

Inferences from the action of purgatives like magnesium sulphate.

The marked loss of water in the large intestine; equilibrium in the small gut.

Abnormal: Traumatic fistulæ; alteration in the reaction of saliva in pregnancy (Beers), and certain diseased states; dyspepsia; its multitudinous forms; constipation, diarrhœa, dysentery; intestinal obstruction.

DEMONSTRATIONS.

Comparative: The stomach of a *frog, bird, ruminant, carnivore,* etc.

The action of saliva on starch.

Rapidity of the action.

Destruction of digestive power by boiling.

Extracts of the mucous membrane of the pig's stomach.

Artificial digestion with the same, of fibrin, etc.

Influence of: reaction; temperature, etc.

Character of the products formed.

Pancreatic extracts:

Artificial pancreatic digestion of proteids, starch, fats.

Influence of varying conditions.

Tendency to putrefaction. Comparison with artificial gastric digestion in this respect.

Formation of leucin and tyrosin.

Bile : Reaction, color, absence of albumin, presence of mucus.

Bile pigments and cholesterin prepared from gall stones.

The bile salts and bile acids.

Test for bile pigments (Gmelin),

Test for bile acids (Pettenkofer).

Emulsifying power of bile.

Precipitation of gastric digestive proteids by bile, (or bile salts).

Peristaltic movement of the intestines.

Naturally injected lacteals.

X. Excretory Processes.

No sharp line of demarcation between *secretion* and *excretion*.

Excretions are useless or harmful; secretions serve some further special use in the economy.

Similar physical and chemical laws apply to both.

PHYSIOLOGY OF THE SKIN AS A SECRETING ORGAN.

The skin is: 1. Protective. 2. Secretory. 3. A regulator of temperature, blood density, etc. 4. Seat of various sense organs.

Histological and *physiological*:

Epidermis (cuticle); *dermis* (cutis vera, corium).

Transitional forms and stratification of epithelial cells; pigmented layer, determining color of skin in races and individuals.

Modifications of the epidermis: hair, nails, feathers, scales, etc.

Muscular and elastic tissue of the *corium;* their functional importance.

Blood supply of the skin.

Glands: 1. *Sudoriparous:* simple tubular; coiled at origin; spiral near termination (adaptation to stretching).

2. *Sebaceous:* saccular; open into mouths of hair follicles; oil glands of birds.

Hairs: Papilla; hair follicle; muscles.

SWEAT AND ITS SECRETION.

Sweating removes waste matters and lowers temperature.

Amount of the perspiration determined by: moisture and temperature of the air; exercise; nature of food, etc.

Method of determination of the quantity of sweat for a given period.

Sensible and insensible perspiration.

Reaction of sweat is acid usually (fatty acids); that of the pure secretion of the sweat glands, alkaline.

Sweat contains about 2 per cent. of solids.

Composition; water; sodium chloride; several fatty acids; neutral fats and cholesterin; trace of urea.

The smell of sweat due to volatile fatty acids.

The secretion of the sebaceous glands; function.

Respiratory function of the skin.

CO_2: 4-10 grams in 24 hours.

Importance of this function in *Amphibia*.

The varnished rabbit.

Evaporation from the skin.

Abnormal: Excretion of drugs and poisons by the skin.

Vicarious action of the skin.

Skin diseases; sebaceous cysts; comedones.

"Critical" sweatings.

Recent Investigation: Electric phenomena of secretion (Baylis and Bradford).

NERVOUS MECHANISM OF SWEAT SECRETION.

Sweating may be and usually is accompanied by increased blood supply to the skin; but such is not essential.

Sweating from fear with pallid skin.

Experimental: sweating in the soles of the feat of the cat, on stimulation of the sciatic nerve, after ligature of the blood vessels; after amputation of the limb.

Sweating on stimulation of the central end of the divided sciatic, the opposite nerve being intact.

No secretion after administration of atropin.

Conclusions: 1. Sweating is effected by the protoplasmic activity of the gland cells, under the contro of the nervous system. 2. There is a reflex nervous sweating mechanism.

Analogies with the submaxillary gland.

Course of the several fibres.

Sweat centres.

Action of drugs: local and central.

Absorption by the Skin.

In the frog and other amphibians of considerable importance.

Insignificant in man.

Absorption of drugs.

PHYSIOLOGY OF URINARY SECRETION.

Anatomical: The simplest vertebrate kidney takes

the form of a tubule with special arrangement of the blood vessels and epithelial lining.

The *water-vascular (nephridia)* system of invertebrates.

Mammalian kidney:

Lies in portective fat; its capsule readily removeable; hilus; cortex; medulla; pyramids of malpighi; papillæ projecting into *calices* of *pelvis*.

Main vessels and nerves.

The histology of the kidney in minute detail.

The kidney is a mass of densely aggregated secreting tubes and blood-vessels held together by a little connective tissue.

These tubes are *secreting* and *collective;* and furnished with epithelium of greatly varying form in different parts.

The blood supply of the kidney is abundant and peculiar.

PHYSICAL AND CHEMICAL PROPERTIES OF URINE.

Reaction in man acid owing to NaH_2PO_4 and KH_2PO.

Color, odor, etc.

Specific gravity of the urine collected during 24 hours: 1020.

Constituents:

1. *Water.*
2. *Inorganic salts* in solution; most abundant $NaCl$.

In smaller quantity: $CaCl_2$, Na_2SO_4, K_2SO_4, Na_2HPO_4, NaH_2PO_4, K_2HPO_4, H_2KPO_4, $Ca_3(PO_4)_2$, $Mg_3(PO_4)_2$ etc.

Source of phosphates: food; metabolism.

Carbonates exist in but small quantity.

3. *Nitrogenous-crystalline bodies:*

Derived from bodily metabolism.

(*a*) Urea, uric acid.

(*b*) Less oxidized forms of proteid metabolism: kreatinin, xanthin, hypoxanthin.

(*c*) Hippuric acid, ammonium oxalate.

4. *Non-nitrogenous:*

Organic acids: lactic, succinic, formic, oxalic, phenylic, etc.

5. *Pigments:*

Special pigments: urobilin, purpurin, indican.

Abnormal: Alkaline reaction of *cystitis*.

Presence of blood, pus, epithelium, spermatozoa.

Oxalic acid diathesis.

Recent investigation: Experimental examination of the secretion of oxalic acid (dog) under a varying diet (Mills).

Comparative: reaction of urine of *carnivora* is strongly acid; of *herbivora*, alkaline.

Abundance of phosphates and sulphates in the urine of carnivora; of carbonates in that of herbivora.

Abundance of *uric acid* in the urine of reptiles and birds.

Recent investigation. The urine of chelonians (Mills).

Amounts of certain constituents of urine passed in 24 hours (Parkes).

Urea	33.180 grams.
Uric acid	.555
Phosphoric acid	3.164
Chlorine	7.000
Sodium	11.090
Total solids	72.000

Variations in the several constituents of urine; on what they depend.

Urine becomes alkaline on standing in a warm atmosphere.

$$CO \begin{cases} NH_2 \\ NH_2 \end{cases} + 2H_2O = CO_3(NH_4)_2$$

Physical and chemical characters of urea and uric acid.

THE SECRETORY PROCESS IN THE KIDNEY.

New methods of investigation: the oncometer, and oncograph.

The kidney curve.

Influence of blood pressure on the secretion of urine.

The pressure in the glomeruli is the final determining factor.

Variations of blood pressure effected experimentally and their influence on renal secretion.

Complementary action of kidney and skin.

Effect on the secretion of urine, of the application of cold and heat to the skin.

Effect of injection of water into the blood.

THE PART TAKEN BY THE RENAL EPITHELIUM.

The double renal blood supply in *amphibians*.

Experiments of Nussbaum (ligatures).

Experiments of Heidenhain (injections).

Albuminous urine after ligature of the vessels of the kidney.

Recent investigations.

Theories of urinary secretion (Bowman, Ludwig, Heidenhain, Nussbaum).

We may conclude that the cells of the kidney possess, by virtue of their own peculiar constitution, the power to secrete all the constituents of urine.

MICTURITION.

Anatomical and physiological. Structure of the bladder; unstriated muscle cells; their arrangement; mucous membrane provided with mucous glands. The sphincter of the bladder.

The *ureters*: muscle cells; mode of opening into the bladder.

The urethra is a closed elastic tube.

The secretion of urine in **constant**.

Micturition may take place as an involuntary reflex **act, but usually it is** *initiated* **by a volition.**

The determining cause of the reflex seems to be the quantity (and quality) of urine in the bladder.

The centre has been experimentally demonstrated in the dog to be in the lumbar cord (Goltz).

Tonic muscular contraction of the *sphincter vesicæ*.

Abnormal: Involuntary micturition in diseases of the spinal cord; dribbling of urine; nocturnal incontinence of urine; action of the bladder under urethral obstruction.

DEMONSTRATIONS:

The properties of urine freshly voided.
Urine of herbivora and carnivora.
Urine as modified by standing; sediments, etc.
Quantitative estimation of sugar, urea, etc.
Separation of uric acid and urea.
Specimens of abnormal urine.

XI. Metabolism: Nutrition.

The chain of transformations, taking place in the food between its entrance into the body and its leaving it as waste-products, is included under the term metabolism.

Metabolism therefore deals with chemical changes and is governed by chemical laws.

The energy *latent* in the food is set free by the metabolic changes of the body, and reappears as kinetic energy and heat.

The series of changes between food and waste-products involves processes both analytical and synthetical.

The total metabolism of an organ or of the body is but the sum of the metabolism of its component cells.

The metabolism of amœba.

GLYCOGEN.

The liver is the largest gland of the body; preponderance in fœtal life; abundant blood supply.

Wide distribution of glycogen (animal starch) in the animal kingdom.

Formation of glycogen as influenced by the *kind* of food.

Fat cannot form glycogen.

Disappearance during continued starvation.

Conversion into sugar after death.

Theories of glycogen formation.

Glycogen is probably a reserve of carbohydrates to be changed into sugar as required.

Analogy with plants.

DIABETES.

Clinical and artificial.

How the latter is produced.

The diabetic area of the Medulla is related to the vaso-motor centre.

Vaso-motor changes in the liver after puncture of the diabetic area.

Diabetes is due to changes in the glycogenic function of the liver originated by the nervous system; but the *modus operandi* is obscure.

Clinical diabetes influenced by the nature of the food.

FAT FORMATION.

Adipose tissue: forming and formed; its fluctuation.

Pathological fatty degeneration of various forms of protoplasm.

The "ripening" of cheese.

Fattening foods: carbohydrates, hydrocarbons.

Fat formation from the fatty acids of the alimentary canal.

Proteids as a source of fat.

Experiments of Lawes and Gilbert:

For every 100 parts of fat of the food of a fattening pig, 472 parts fat laid up in the body of the animal.

Butter of the cow, wax of bees, etc., out of proportion to the fatty food taken.

The fat of the animal does not correspond in composition with the fat eaten.

Theories of the formation of fat from proteids, carbohydrates, fatty acid, leucin, (amido-caproic acid) etc.

THE METABOLISM OF THE MAMMARY GLAND.

Histological: A compound racemose gland; its cells: short columnar.

Composition of milk:

Reaction alkaline: may be acid even in the gland.

Solids:
- Proteids:
  - Casein.
  - Serum-albumin.
- Carbohydrates: Milk sugar (lactose).
- Fats.
- Salts: phosphates of Ca, Mg, K; K Cl.

The *solids* constitute about 10 per cent. of milk.

Milk is an emulsion.

The casein is "*particulate*," but may be filtered out through porous earthenware.

Cow's milk contains 3-5 per cent. of casein.

Human milk about half that proportion.

Colostrum is deficient in casein and rich in albumin.

Colostrum corpuscles.

The fatty globules of milk are surrounded by albuminous envelopes.

The natural curdling of milk is owing to the conversion of its milk sugar into *lactic acid*, in the presence of micro-organisms.

Comparison of milk and blood coagulation.

Butter is a mixture of many fats, but consists principally of palmitin, olein, stearin.

DEMONSTRATIONS.

The properties of milk.

The proteids, fat and carbohydrates of milk are all the outcome of the protoplasmic activity of the gland cells.

Experimental feeding.

Histological investigation.

Influence of the nervous system, drugs, etc.

SPLENIC METABOLISM.

Anatomtical and physiological. Is a modified lymphoid gland.

Elastic capsule; trabeculæ; stroma; pulp: containing delicate connective tissue, capillaries, malpighian bodies; corpuscles: splenic; lymphoid; red.

Muscle cells of capsule and trabeculæ.

Extirpation of spleen; possible compensatory changes.

Variations in size during digestion.

Rhythmic variations.

Curve of these variations.

Comparison with kidney curve.

The spleen may be contracted *directly* by stimulation of the splanchnic or vagus nerves; by mechanical excitation; and reflexly by the stimulation of a sensory nerve; also by stimulation of the medulla.

Metabolism and special chemistry.

Ferruginous proteid.

Pigments rich in carbon.

Large amount of sodium salts and phosphates, and small amount of potassium salts and chlorides, in the ash.

Abundance of extractives: acetic, formic, butyric, succinic, and lactic acids; inosit; leucin; xanthin; hypoxanthin and uric acid.

Resemblance in metabolism of other blood-forming glands.

The active metabolism seems to be associated with the digestive process.

Abnormal: Enlarged spleen of fevers; "ague cake."

UREA AND ALLIED BODIES.

Urea represents almost the entire nitrogenous waste of the body.

Absence of urea in muscle.

The nitrogenous bodies, kreatin, xanthin and hypoxanthin arise from muscular metabolism.

The kreatinin of the urine is not derived from the kreatin of muscle.

The greater part of the urea of the urine is simply *selected* from the blood by the kidney cells.

It is probable the liver cells convert into urea the *leucin* of pancreatic digestion.

The amount of leucin (and tyrosin) formed in the alimentary canal seems to be proportional to the excess of proteid of food.

Urea is found in the liver.

Theory of urea formation in the liver.

In certain diseases of the liver, leucin and tyrosin appear in the urine.

Uric acid does not seem to give rise to urea.

General conclusions: 1. The antecedents of urea in the blood are kreatin and other nitrogenous crystalline products of muscle and other metabolism, together with leucin and similar bodies.

2. The liver, and possibly the spleen effect the transformation of these bodies into urea.

3. Possibly the kidney transforms nitrogenous crystalline bodies into urea.

These conclusions are only probable and not demonstrated.

URIC AND HIPPURIC ACIDS.

Uric acid is a less oxidized product than urea, but not a demonstrated antecedent.

Uric acid replaces urea in reptiles and birds.

Hippuric acid replaces uric acid in the urine of *herbivora*.

Formation of Hippuric acid:
Benzoic acid. Glycocine.   Hippuric acid:

$$C_6H_5.CO.OH + \left.\begin{array}{l}H_2N\\HO_2C\end{array}\right\} CH_2 = C_6H_5.CONH.CH_2.CO_2H + H_2O$$

or   or
Glycine.   Benzoyl-amido-acetic acid.

(after Remsen).

The transformation appears to take place in the kidneys, and in some animals, in the liver.

Abnormal: The excessive formation of urea and uric acid in certain diseased states (Diabetes, gout, etc).

STATISCAL EXAMINATION OF NUTRITION.

Relative proportions of the several tissues of the body.

Relative loss in weight during a starvation period.

Nitrogenous excretion during the fasting period.

Luxus consumption (Bidder & Schmidt), "Floating capital."

Normal diet: how framed.

Ranke's dietary:	Proteids.	100
	Fat.	100
	Amyloids.	240
	Salts.	25
	Water.	2600

STUDY BY COMPARISON OF INCOME AND OUT-PUT.

Comparison of *ingesta* and *excreta.*

Income: Food *minus* the fæces.

Output: Respiratory products.
Perspiration.
Urine.

Nitrogenous equilibrium.

Determination of factors in a given case of experimental feeding.

Constructive and destructive metabolism.

"Tissue proteids" or "morphotic protein;" "circulating proteids." (Voit).

Urea derived from both.

Proteid food increases both nitrogenous and non-nitrogenous metabolism.

Carbohydrates may probably be *directly* converted into fat.

Other view: carbohydrates shelter the fats already stored up from oxidative changes.

In *herbivora* a larger proportion of the O consumed reappears as CO_2 than in *carnivora*, and in the latter more as water.

Carbohydrate.
$$C_6H_{10}O_5 + 6O_2 = 6CO_2 + 5H_2O.$$
Tri-stearin.
$$C_{57}H_{110}O_6 + O_{163} = 57CO_2 + 55H_2O.$$

Fats have more potential energy than carbohydrates weight for weight.

Gelatine cannot wholly replace proteids in food; but it may render the same quantity of proteid more effective.

We may suppose gelatine capable of rapid decomposition into urea and a fatty portion.

Water, when increased, augments the quantity of urea excreted.

Salts are regulators of metabolism.

Practical. The regulation of diets so as to combine cheapness and physiological fitness.

THE STUDY OF BODILY ENERGY.

"The animal body is a machine for converting potential into actual energy" (Foster).

The energy available for the body depends on the difference of *chemical potential* between the food and the waste-products.

In estimating this the *final* result of chemical decomposition, not the intermediate stages, is considered.

Instability of the molecule is associated with absorption of energy and *vice versa*.

Heat and mechanical labor represent the totals of the processes by which energy is set free by metabolism.

SOURCES OF ENERGY.

The main source of all the energy of the body is from the oxidation of food.

On the one hand we have proteids, fats, carbohydrates; on the other their oxidation products as CO_2, H_2O, CO $\begin{cases} NH_2 \\ NH_2 \end{cases}$

How the potential energy of food is calculated.
How available energy is estimated.
Calculation of the available energy of proteid :

	Gram degrees.	Kilometres.
1 gram proteid	5103	2161
⅓ gram urea	735	311
Available energy of proteid	4368	1850

Relations of function, heat, and movement.

EXPENDITURE OF ENERGY.

Liebig's division of foods : respiratory or non-nitrogenous ; plastic or nitrogenous.

The urea of the urine is not appreciably increased by muscular exercise.

The nitrogenous part of the molecule of muscle is therefore, probably, not decomposed by muscular exercise.

The amount of CO_2 excreted is increased many times by muscular labor.

The experiments of Fick & Wislicenus on muscular energy.

Conclusion : The energy of muscle is not derived from proteid metabolism exclusively.

ANIMAL HEAT.

Heat is the outcome of the oxidative processes of the body.

These processes take place in the *tissues*.

Heat in a *resting* animal is the sole representative of the difference of energy between food and waste-products.

The muscles and glands (especially the liver) are the chief sources of heat.

Heat is lost chiefly by the skin and lungs.

Poikilothermal (cold-blooded), homoiothermal (warm-blooded) and hybernating animals compared as to temperature.

Variations in the temperature of different parts of the body.

The liver is the warmest organ.

The blood is the great medium of heat distribution.

Abnormal: lower temperature of persons with patent *foramen ovale*.

Methods of determining the bodily temperature.

Differences for axilla, mouth, rectum, etc.

Examples: temperature of man   37.60 C.
swallow   44.00.
wolf   35.24.

Variations with age, sex, muscular exertion, ingestion of food, etc.

Daily variations: *maximum* from 9 a.m to 6 p.m; *minimum* from 11 p.m to 3 a.m.

Difference about 1°.

REGULATION OF TEMPERATURE.

1. By *variations in loss*, owing to warming fæces and urine; warming expired air; evaporation of the water of respiration; in radiation, conduction and evaporation by the skin.

About three-quarters of the total loss takes place through the skin.

There is no pulmonary mechanism for heat production.

Right ventricle is warmer than the left.

The skin is the chief regulator of the loss of heat.

Its sweat glands; its vascular supply.

Tolerance of very high temperatures.

Increase of blood temperature with the application of cold externally.

2. By *variation in heat production.*

Cold stimulates to increased oxidation in homoiothermal animals and the reverse in poikilothermal.

A curarized homoiothermal animal behaves like a poikilothermal one.

Division of the spinal cord is followed by lowered temperature.

Conclusions: 1. The production of heat is under the control of the nervous system.

2. The mechanism is reflex.

3. There is a thermogenic centre (in the pons varolii ?)

4. The centre is probably double, *i e.*, a heat producing and a heat inhibitory division or centre.

Pathological: Fever, in which there is increased metabolism.

Heat of inflammation; death from burns; death from cold.

Influence of the nervous system over nutritive processes.

Bearings on this subject of diabetes; paralytic secretion of saliva.

"Trophic" nerves.

Pathological : acute bedsores ; degeneration of muscle in diseases of the spinal cord; inflammation of the eye after section of the 5th nerve ; pneumonia after section of the vagi.

Demonstrations.

The extraction of glycogen from the living tissue of the liver and muscles.

Conversion into sugar.

XII. The Nervous Centres.

The great nervous centres are the spinal cord and brain.

THE SPINAL CORD.

Anatomical and physiological:
Coverings: *Dura mater;* protective and supporting; *arachnoid* non-vascular; *pia mater* vascular.

Cauda equina; filum terminale (of gray matter).

Fissures: Anterior median, posterior median.

Cerebro-spinal fluid.

Central canal lined with ciliated columnar epithelium.

Nerve-roots: Anterior and posterior. Line of entrance defines the *lateral fissures.*

The gray matter of the cord forms a central crescent around which the white matter is placed.

The cord histologically consists of nerve **cells,** *neuroglia,* and nerve *fibres.*

Two lateral halves of cord united by a gray (posterior) and a white (anterior) commisure.

The gray matter:
Anterior and posterior *cornua.*

Histological: Nerve cells; their interlacing processes; nerve fibres.

Special collections of cells: (*a*) Multipolar cells of anterior cornu. (*b*) Clarke's column: large round cells at the base of posterior cornu in middle *dorsal* region. (*c*) Cells of lateral cornu (*intermedio-lateral tract*).

The Course of the nerve roots.

Fibres of anterior roots pass:
(*a*) Directly to nerve cells of anterior cornu.
(*b*) To posterior cornu through gray matter.
(*c*) To lateral white column of same side.
(*d*) To the anterior cornu of the opposite side through the isthmus.

Fibres of the posterior roots pass:
(*a*) Chiefly into posterior white columns, and after a short course into the gray matter, some connecting with its cells, others passing to the opposite side of the cord.
(*b*) Some to the posterior cornu directly and are likely connected with scattered cells of that region.

Substantia gelatinosa of Rolando.

Fibres of the white matter (*fibre tracts*):

These are medullated; those of the gray matter non-medullated.

Anterior column { Antero-lateral column. Direct pyramidal tract or antero-median: column of Türck.

Lateral column. { Crossed pyramidal tract. Direct cerebellar tract (cervical and dorsal regions).

Posterior column. { Cuneate fasciculus (Burdach's column). Posterior medium column or tract of Goll, distinct above middle of dorsal region.

The size of the fibres and its significance.

Smallest in *posterior* columns (posterior median).

Largest in crossed pyramidal tract: the path of motor impulses.

Large fibres of anterior column run in the direct pyramidal tract.

Comparison of relative proportion of gray and white matter in the different regions of the cord.

Cervical and lumbar enlargements.

REFLEX FUNCTIONS OF THE SPINAL CORD.

A reflex action cannot be automatic, but depends essentially upon a *stimulus* reaching the centre from without it.

This stimulus usually acts upon *end-organs*.

Latent period of action of the *cells* of reflex centre.

"Summation" of impulses.

Modification of impulses.

Variations in the reflex according to: (*a*) The intensity of the stimulus. (*b*) Site of application of the stimulus. (*c*) Condition of cord (strychnia).

Inhibition of reflexes by centres in the brain and the spinal cord.

Excitation of reflexes through the organs of special sense.

Reflex time.

Economy of cerebral energy owing to the increased facility of discharge along certain lines in the spinal cord from repeated action (habits).

The spinal cord is neither conscious nor intelligent, and though reflexes usually result for the benefit of the animal they may, in certain cases, lead to its destruction.

The co-ordination of a race and of an individual.

THE CORD AS A COLLECTION OF CENTRES OF AUTOMATIC ACTION.

Irregular automatism (volition) is wanting in the cord.

The behavior of a decapitated frog when left wholly to itself.

The rhythmic movements of animals with divided spinal cord.

Automatic centres in the cord: micturition, defœcation, erection of penis, parturition, etc.

Theory of reflex action in maintaining the condition natural to muscles when passive *versus* that of "tonic" action.

"Tendon phenomenon."

It is certain that the *nutrition* and *irritability* of the muscles depend on the integrity of the spinal cord and the nerves connecting it with the muscles.

THE CORD AS A CONDUCTOR OF IMPULSES.

Methods of investigations: 1. Anatomy. 2. Comparative anatomy and physiology. 3. Embryology (Flechsig). 4. Experiment. 5. Pathology and clinical medicine. Degeneration of fibres (Waller).

Afferent impulses along nerves not cranial reach the brain by passing through the spinal cord; similarly the cord is the great highway along which volitional impulses travel.

An impulse is not conveyed to the brain along the identical fibre in which it began its course. The size and shape of the cord do not admit of such a conception of conduction.

Diagrams of sectional variations of the gray and white matter of the spinal cord.

The part taken by the nerve centres of the cord in conduction : "relay stations."

Decussation:

All impulses decussate somewhere in the nervous centres between the point at which they enter or originate and the point at which they leave them.

The decussation of impulses whether sensory or motor is almost complete at about the junction of the medulla oblongata and pons varolii.

Sensory impulses cross lower in the spinal cord and either directly or soon after entering it.

Motor impulses cross chiefly in the medulla.

The crossed pyramidal tract

The direct pyramidal tract.

The direct cerebellar tract.

Volitional impulses travel in the *lateral* columns, become connected with the gray matter of the *anterior cornua*, and thence pass by the *anterior roots* along the efferent nerves to the muscles.

Sensory impulses pass to the *posterior cornua* by way of the *posterior columns*, cross soon to the opposite side, and thence ascend to the brain by the *lateral* columns.

Tracts of degeneration.

Experimental:

Results of section of different parts of the cord (Yeo's physiology).

The functions lost by partial sections of the cord, not including the whole or greater portion of gray matter, may be finally recovered without the reunion of divided parts.

Impulses may select unusual paths when the ordinary tracts are not intact.

Tactile and sensory impulses probably pass along different tracts.

Respiratory and vaso-motor impulses seem to travel by the *lateral* columns.

The spinal cord as a whole is irritable.

The white matter is irritable.

The gray matter is kinesodic and æsthodic.

THE BRAIN.

General anatomy of the brain of a mammal; of man.

Comparative: Brain of *fish, frog, bird* and numerous *mammals* compared.

The nervous system of the *Coelenterates, asteroids annelids, mollusks, arthropods* etc., compared.

The brain as a whole may be considered as a mass of cells and fibres with supporting tissue (neuroglia), having arrangements allowing of economy of space; with the gray matter (cells) arranged in definite masses; and the whole so anatomically and physiologically connected as to permit of co-ordination of functions.

THE MEDULLA OBLONGATA.

The medulla is the seat of the following centres:
Respiratory.
Vaso-motor.
Cardio-inhibitory.
Deglutition.
Secretion of saliva.
Vomiting.
And according to some authors,
Centre for suction.
Mastication.
Closure of eyelids, etc.

These centres preside over the "organic" functions of the body.

The medulla is one of the great centres for the co-ordination of the muscular movements of the body.

No animal can survive the removal of its medulla for any considerable period of time; a mammal dies instantly.

THE CEREBRUM.

Histological: The minute structure of a cerebral convolution as seen in a vertical microscopic section.

Various kinds of *cells; fibres; neuroglia.*

EXPERIMENTAL REMOVAL OF THE CEREBRUM.

"*Shock:*" abolition of the function of a nervous centre.

Some animals bear removal of the cerebrum better than others (batrachians, reptiles, birds).

Psychical activity seems to be abolished.

The animal becomes a machine worked by reflexes.

Condition of a frog and a pigeon compared after the removal of the cerebrum.

There is loss of perception; of voluntary movement and of will power generally.

There is retention of the power of swallowing, etc., of muscular co-ordination; of avoiding obstacles.

The *sensations* remain; the *perceptions* are lost.

Similar but less defined results in a mammal.

THE MECHANISM OF CO-ORDINATED MOVEMENTS.

The semicircular canals:

Relative position; bony and membranous; *ampullæ;* their nerve terminations, etc.; communications; *endolymph.*

Section of the canals in a pigeon:

Dependence of the results on the number of canals divided; relations to recovery.

Similar results in the mammal.

Dizziness in the human subject.

The equilibrium sense.

Dependence on numerous afferent impulses.

Complicated co-ordinations involved.

Theory in regard to the changes in the *pressure* of the *endolymph* within the membranous simicircular canals.

Abnormal: Ménière's disease; vertigo with deafness.

Comparative: Experimental section of the semicircular canals of the shark (Sewall).

FORCED MOVEMENTS.

Rolling; "circus" movements; circular movements; tumbling.

Follow unilateral section of the crura cerebri; of the **pons** varolii; lesion of the medulla, corpora quadrigemina, nates, corpora striata, optic thalami.

These movements may be witnessed when there is no spasm or muscular paralysis.

Probably due to derangement of the co-ordinating mechanism dependent in turn on disordered sensations.

FUNCTIONS OF THE CEREBRAL CONVOLUTIONS.

The cerebrum is, as a whole, the centre of psychical activity: intellection, volition, emotion.

The brain substance is *insensitive*.

The cerebral localization hypothesis; definite functions for circumscribed anatomical regions of the cortex.

The ablation experiments of Flourens.

The teaching of clinical experience.

Aphasia, as associated with lesion of the posterior part of the third left frontal convolution.

Stimulation experiments (Hitzig, Fritsch, Ferrier).

"Motor" and "sensory" centres.

Tactile reflex theory (Schiff).

'Visual area;" "absolute blindness;" "psychical blindness" (Munk).

Theory of "vicarious" action.

Experiments of Goltz on the dog.

Conclusions of Goltz:

Removal of portions of the cerebral cortex leads to psychical impairment; to loss of sensation, and of will power; these are all proportionate to the amount of loss of the convolutions; the kind of functional impairment is independent of the seat of the operative lesion.

Especially does Goltz maintain that there is no paralysis in the dog following the removal of the great motor area of Ferrier and others.

Results of the stimulation of the intact cortex and of the white matter beneath compared.

Variations in the results of stimulation of the cortex dependent on different degrees of the action of morphia, etc.

Cerebral inhibition as modified by drugs, peripheral stimulation, etc. (Heidenhain).

Remarkable increase in the facility with which reflex action may be excited in animals, in which portions of the cerebral cortex have been removed.

The force and duration of such reflexes.

Abnormal: Cortical epilepsy.

Localization of tumours.

THE CORPORA STRIATA AND OPTIC THALAMI.

These are the " basal ganglia " of the cerebrum.

They abound in gray matter (cells).

Some of the fibres of the *crura cerebri* pass to them, and after connecting with their cells ascend to the cortex; others directly through them or past them without such connection.

The lower or anterior fibres of the *crus (crusta)* pass to the corpora striata; the upper or posterior fibres (*tegmentum*) to the optic thalami.

Pathological and clinical.

Lesions involving the basal ganglia of one side of the brain are associated with loss of sensation and voluntary movement of the *opposite* side of the body.

Experimental lesion is followed by the same results.

Lesion of *one* corpus striatum has been in some cases attended with *anæsthesia* of the opposite side of the body.

It may be said provisionally that the *corpora striata* are concerned with the elaboration of *motor* and the *optic thalami* of *sensory* impulses.

When the *optic thalami* are injured, blindness, more or less complete according to the extent of the lesion, follows.

THE CORPORA QUADRIGEMINA.

These ganglia in the mammal correspond to the corpora bigemina or optic lobes of lower vertebrates.

The *anterior pair* of the corpora quadrigemina are alone connected with the *optic tract* and with the *corpus geniculatum* and *optic thalamus*.

Stimulation of the corpora quadrigemina leads to movements of the eyeballs and variations in the size of the pupil.

When the visual axes converge, the pupils contract, and *vice versâ*.

The *centre* for these movements really lies in the front part of the floor of the aqueduct of Sylvius.

Unilateral destruction of the corpora quadrigemina (or optic lobes) is followed by blindness of the opposite eye.

It is inferred that visual impulses are transformed into visual *sensations* in these ganglia.

For a clear visual *perception* the cerebral cortex is required.

" The processes constituting distinct and perfect vision, in fact, begin in the retina, are partially elaborated in the corpora quadrigemina, and further developed in the optic thalami, but do not become perfected until the cerebral convolutions have been called into operation. " (Foster)

Decussation of visual impulses is complete in some animals (rabbit), incomplete in others (dog).

It is incomplete in man.

Abnormal : Hemiopia.

Experiment seems to show that the corpora quadrigemina have some connection with the centres regulating blood pressure, respiration, etc.

THE CEREBELLUM.

Anatomical: The arrangement of the parts as seen in a vertical microscopic section. The cells of Purkinje and their connections.

Experimental :

Lesions of the cerebellum cause loss of co-ordination, in proportion to their extent.

Unilateral section of the cerebellum (middle peduncle) causes forced movements and *nystagmus*.

There is no evidence that the cerebellum is connected with the sexual functions.

The sexual centres are situated in the lumbar cord.

Abnormal: Peculiar gait in diseases of the cerebellum.

CRURA CEREBRI AND PONS VAROLII.

They contain an abundance of gray matter.

The crura are physiological highways between the spinal cord and the higher part of the brain.

Both the crura cerebri and pons varolii are engaged in co-ordination.

Forced and disordered movements result from their section.

Partial decussation of impulses in the pons.

Abnormal: Unilateral disease involving the pons, causing paralysis of the face on the same side.

CEREBRAL TIME.

"Reaction time."

"Personal equation."

Reaction time for feeling $\frac{1}{7}$; for hearing $\frac{1}{6}$; for sight $\frac{1}{5}$ of a second.

"Reduced reaction period" about $\frac{1}{10}$ of a second.

Volition time from $\frac{1}{5}$ to $\frac{1}{30}$ of a second.

CIRCULATION OF THE BRAIN.

Peculiarities of the blood vascular supply, arterial and venous.

Limited anastomoses.

Vaso-motor mechanism.

Cerebro-spinal fluid.

Abnormal: Compression from effused blood.

NOTE.—For the special physiology of the cranial nerves the student may consult the text books of Foster, Yeo, and Dalton. The details would occupy too much space here, and general principles have been already considered.

The same remarks apply to the sympathetic system of nerves

Recent investigation on nerves:
The functional union of a motor and a sensory nerve after section (Reichert).

DEMONSTRATIONS.

A frog or pigeon after removal of the cerebrum.

A frog after removal of all the parts of the brain except the medulla and cerebellum.

Comparison between a frog with a spinal cord only; one with nervous centres intact; and one in which the cerebrum has been removed.

The croaking frog; the balancing frog.

XIII. The Physiology of Vision.

Anatomical :
The compound eye : the simple eye.

The gradual development of the eye through lower animal forms. Functional variations corresponding with anatomical differences.

Dissection of eyes of the ox: in the natural state; after freezing or boiling.

The muscles of the eyeball.

The *optic* nerve, entering obliquely.

The coats of the eye : The *sclerotic*, strong and protective ; the *choroid*, pigmented and vascular ; the *retina*, delicate nervous tissue.

In a certain sense all the rest of the eye exists for the retina.

The *cornea*, a section of a lesser sphere set into a section of a greater (sclerotic).

Bowman's membrane (connective tissue).

Membrane of Descemet (elastic).

The cornea has no blood vessels of its own ; is nourished by diffusion.

Ciliary muscle, attached at corneo-sclerotic junction; composed of involuntary muscle cells; inserted into choroid.

Ciliary processes of the choroid : highly vascular folds.

The *Iris*: is continuous with the choroid, very vascular; connective tissue; radiating non-striated muscular cells (dilator) and circular muscle (sphincter); *uvea*.

The Retina: Its 7-8 layers.

The large ganglionic cells of the second layer connect on the inner side with axis nerve cylinders, and on the outer with the granules and through them with the inner ends of the rods and cones.

The *histology* of the rods and cones in minute detail.

The *macula lutea* and *fovea centralis*.

These lie in the visual axis.

Anatomical characteristics of the yellow spot:
1. Greater thickness (except at *fovea*).
2. Large number of *ganglion cells*, all distinctly bipolar.
4. Large number of *cones* as compared with rods.
4. No rods in fovea; cones very long and slender.
5. Middle of fovea *thinnest* part of retina.

(After Schäfer.)

Few blood vessels in the retina.

Ora serrata; *canal* of *schlemm*.

The *suspensory ligament*: its relation to the *hyaloid membrane, zonule of zinn* and *canal of Petit*.

The *crystalline lens*: shape in different animals.

Its *laminæ*; *capsule*.

The *anterior* and *posterior* chambers.

Aqueous and vitreous humours.

The distribution of *blood-vessels* in the eye.

The *pupil* is not a structure but the central opening of the iris.

Appendages of the eye:
Eyelids; meibomian glands; conjunctiva; nictitating membrane of birds, etc.
Lacrymal gland; lacrymal duct.

GENERAL PHYSIOLOGY.

The eye is an arrangement of focussing bodies, protected by coverings, with a window for the admission of light; a curtain regulating the quantity admitted; a sensitive membrane for the formation of the images; surfaces for the absorption of superfluous light; apparatus for the protection of the eye as a whole, and for keeping exposed parts moist and clean.

For vision there is provided: (a) A peripheral *sense organ*. (b) An afferent nerve. (c) A nervous centre.

Conscious perception of an object is associated with the cerebrum.

Sensations and *judgments* compared.

THE EYE AS AN OPTICAL INSTRUMENT: DIOPTRICS.

Light: vibration of ether; physical and physiological aspects.

The eye as a *camera obscura*.

Perception of light and perception of an image compared.

The purpose of the dioptric apparatus is the *formation* of *images* on the retina.

The simplest dioptric mechanism must consist of *two media* with an intermediate curved *surface*.

The optical properties depend on: 1. The curvature of the surface. 2. The relative refracting power of the media.

The *foci* of the refracting media all fall upon the *optic axis*, which latter meets the retina a little above and internal to the *fovea centralis*.

Refraction of the light entering the eyes is effected chiefly at the *anterior* surface of the cornea and the *anterior* and *posterior* surfaces of the *lens*.

Necessary *inversion* of images.

Optically the *cornea* may be considered as having but *one* surface.

The refracting power of the *lens* may be considered *uniform*.

The aqueous and vitreous humours have a refractive power almost equal to that of the cornea.

The "diagrammatic eye."

Its refracting surfaces: 1. Anterior surface of the cornea. 2. Anterior surface of the lens. 3. Posterior surface of the lens.

The *principal posterior focus* is that point at which all rays, falling on the cornea parallel to the optic axis, are brought to a focus.

ACCOMMODATION.

Accommodation is adaptation of the mechanism for exact focussing of light, resulting in clear images on the retina.

Dependence upon *direction* of rays (entering angle) and *curvature* of refractive surface.

Diffusion circles.

Scheiner's experiment.

The near point; the far point.

The *emmetropic* eye: dioptric mechanism normal.

The *myopic* eye: parallel rays focussed anterior to the retina. Long eye.

The *hypermetropic* eye: parallel rays focussed posterior to the retina. Short eye.

The *presbyopic* eye: there is defect of accomodation.

Diagrams illustrating these conditions.

HOW ACCOMMODATION IS ACCOMPLISHED.

The passive eye is accommodated for parallel rays (objects at an infinite distance).

Accommodation for near objects results in:

(*a*) Contraction of the sphincter of the iris leading to *narrowing* of the pupil.

(*b*) Increase in the curvature of the anterior surface of the lens.

The curvature of the cornea is not altered in accommodation.

The lens is rendered more *convex* anteriorly by the action of the *ciliary muscle* in pulling forward the *choroid*, thus relaxing the *suspensory ligament*.

Diagram of the mechanism.

Purkinje's images.

VARIATIONS IN THE SIZE OF THE PUPIL.

The pupil *dilates* with *decrease* of light; *divergence* of the visual axes (distant vision); *contracts* with *increase* of light and *convergence* of the visual axes (near vision).

The muscular mechanism of contraction.

The condition of the eye in sleep.

Nervous supply of the iris:

1. Short ciliary nerves from lenticular (ophthalmic, ciliary) ganglion, connected with *third* nerve.

Long ciliary nerves from nasal branch of ophthalmic division of *fifth* nerve.

2. The cervical sympathetic (through ciliary ganglia).

3. Nasal branch of the ophthalmic division of the *fifth* nerve.

Diagrams showing these relations.

Experimental:

1. *Division* of *cervical sympathetic* followed by *contraction* of the pupil.

2. *Stimulation* of the peripheral end of the same followed by *dilation*.

3. *Division* of the optic nerve is followed by *dilation* of the pupil.

4. Division of *third* nerve with stimulation of the optic nerve is not followed by contraction.

5. *Stimulation* of the peripheral end of the divided *third* nerve is followed by marked contraction.

6. *Stimulation* of the centre is followed by contraction.

The centre is situated in the *front part of the floor of aqueduct of Sylvius*.

7. The *centre* having been *removed* stimulation of the retina is *not* followed by contraction of the pupil.

8. Illumination of one eye is followed by contraction of the pupil of the opposite eye.

Conclusions:

1. The contraction of the pupil is normally a reflex action.

2. The *afferent* nerve is the optic.

The *efferent* the third nerve.

3. The sympathetic is the efferent dilator nerve. Tonic action.

4. There is *associated action* of the centre (or centres) for reflex action (see 8 above).

A local mechanism.

Exhaustion of the cerebral centre (alcohol, dying).

Subordinate centre (*centrum ciliospinale inferius*) in lower cervical and upper dorsal cord.

The *accommodation* centre is situated in the *third ventricle;* connected with the most anterior bundles of the roots of the third nerve.

The nerves : ciliary, through ophthalmic ganglion to *third* nerve.

IMPERFECTIONS OF THE OPTICAL APPARATUS.

The *fovea centralis* is the seat of the most perfect visual sensation.

Spherical aberration of a lens owing to its form.

How counteracted in the actual lens.

Assistance from the iris.

Astigmatism :

Owing to the refracting surfaces not being perfect sections of a sphere. Different *foci*.

Generally traceable to the cornea.

In an eye astigmatic in the vertical meridian horizontal lines are focussed sooner than vertical ones and *vice versâ*.

Hypermetropic and myopic astigmatism.

Regular astigmatism ; irregular astigmatism.

Chromatic aberration :

Arises from the different refrangibility of the several rays of the spectrum ; rays of the violet end are focussed sooner than those of the red end.

Entoptic phenomena :

Muscæ volitantes; effect of tears on cornea ; the imperfection of the lens, margin of pupil, etc.

The refracting surfaces are not truly centered on the optic axis.

VISUAL SENSATIONS.

Sensory impulses originate in end-organs.

Sensation is the result of cerebral action on the impulses.

The *rods* and *cones* are the anatomical elements of the retina sensitive to light.

The optic nerve is not sensitive to light.

The blind spot.

Purkinje's figures.

Diagrams.

Protoplasm is sensitive to light in lowly forms.

Abundance of pigment in the retinal (or choroidal) epithelium.

Retinal purple; "optograms."

Visual purple is *absent* from the cones and fovea centralis; wholly absent from the retinas of some animals.

The yellow pigment of the *macula lutea*.

The green lustre of the eyes of certain animals is due to a special choroidal layer (*tapetum, membrana versicolor*).

It reflects light strongly and gives rise to interference colors.

Electrical phenomena of the eye.

Abnormal: Absence of pigment in eyes of albinos.

THE STIMULUS AS RELATED TO THE SENSATION.

The sensation *outlasts* the stimulus when the latter is of brief duration.

Separate sensations are fused when the succession of the stimuli exceeds a definite rate.

The stronger the light the higher must be the rate of succession.

The duration of the "after image" is in proportion to the strength of the stimulus (light).

"Positive" and "negative" after images.

With gradual increase of the luminosity, the sensations increase; but the increments of the sensations consequent on increments of luminosity gradually diminish.

Lower limit of excitation (threshold); "maximum of excitation." "Range of sensibility" between these limits (Wundt).

The smallest difference of light appreciable is about $\frac{1}{100}$ of the total luminosity.

Weber's law: The increase of stimulus necessary to produce the smallest appreciable increase of sensation bears the same proportion to the whole stimulus.

This law is of general application to the senses.

Example: If 10 grams be placed in the hand it is found that 3.3 grams is the smallest appreciable increment; likewise with 100 grams 33.3 is the smallest appreciable increment.

DISTINCTION AND FUSION OF SENSATIONS.

When images reach a certain degree of retinal proximity they are fused pyschologically (physiological fusion).

Variations in discriminating power in the different parts of the retina.

The sensation areas of the brain are not sharply defined.

Retinal areas.

COLOR SENSATIONS.

Color is determined by the wave length of light.

Spectral colors: *red, orange, yellow, green, blue, violet*. Purple results from the blending of blue and red.

Natural hues may be imitated by the appropriate blending of certain primary color sensations with each other or with white or black.

A color is more or less *saturated* according to the quantity of *white* light mixed with it.

"Pale, rich, deep, bright, subdued, gla g," etc., as applied to color.

Complementary colors are those which when fused (physiologically) produce white.

The fusion of two spectral color sensations may give rise to a different spectral color (*e.g.* red and yellow, producing orange).

All the other spectral colors may, by duly varying the proportions, be derived from the three primary colors, (red, green, blue) with white.

White may be produced by mixing the primary colors.

The retina is most sensitive to blue, least to red.

Theories of color sensations:

1. Hering's chemical theory:

"Fundamental sensations": white, black, red, yellow, green, blue.

These are arranged in pairs so that the presence of one implies the absence of the other.

This is owing to *anabolic* and *metabolic* processes; or to *constructive* metabolism and *destructive* metabolism, which are always related.

The substances are:

Red-green, yellow-blue, black-white.

Constructive metabolism or assimilation.

Destructive metabolism or dissimilation.

When dissimilation is in excess the lighter colors result (red, yellow); when balanced no color.

2. The Young-Helmholtz theory:

All colors are the result of the appropriate mixture of certain primary sensations (red, green, blue or violet).

Diagrammatic representation.

Color blindness:

Red blindness.

Partial color blindness of the peripheral parts of the retina.

Explanation by the above theories.

After images:

Positive and negative.

Colored after images.

The "proper light" of the retina.

Oscillations in color sensations.

VISUAL PERCEPTIONS.

Perception is the final result of a chain of processes and is psychical.

Field of vision.

Localization of sensations.

We cannot localize an object.

Inversion of the image on the retina.

Psychical effects from the use of colored glass.

MODIFIED PERCEPTIONS.

Irradiation: bright objects appear larger than dark ones of equal size.

Contrast intensifies color sensations.

"Filling up" the blind spot.

Visual perceptions independently of light.

Phosphenes.

Ocular spectra (phantoms).

Apparent size:

The eye furnishes the mind with the size of an *image* only.

Judgment of size is a complex process and success depends upon previous experience of a similar kind. Various ways in which the judgment is affected.

BINOCULAR VISION.

"Corresponding" or "identical" points.

Diagram showing them.

Advantages of binocular vision.

The effect of solidity in binocular visions is due to fusion of images which takes place in the cerebrum.

Stereoscopic fusion.

Stereoscopic vision of different colors resulting in rhythmical oscillations of color.

The psychical effects produced by drawings and paintings.

The muscular movements of the eyeballs, and possibly of accommodation, influence our judgments of size and distance.

THE MOVEMENTS OF THE EYEBALLS.

The muscles concerned.

The co-ordinated action of these muscles in elevation, depression, adduction and abduction of the eyeball.

The movements of the eyes in following an object result in keeping the image constantly on "corresponding" points of the retina.

Double images in consequence of departure from the above.

The part of the brain concerned in such movements (see p. 133).

Abnormal: Strabismus.

THE HOROPTER.

The horopter is such a line or surface in the field of vision that the images of the points in it fall on corresponding points of the retina.

When standing upright and gazing at the distant horison the horopter is a plane drawn through the observer's feet.

Diagram.

DEMONSTRATIONS.

Inversion of the image on the retina of the excised eye of an albino rabbit.

Changes in the size of the image according to size and distance of the object.

Effects on the pupil of section of the cervical sympathetic; of stimulation of the peripheral end.

Scheiner's experiment.

Movements of the pupil in the human subject.

Field of vision; field of color vision.

Blind spot.

Purkinje's figures.

Purkinje's images.

Mixture of colors by rotation.

Complementary colors.

After images,

Simultaneous contrast.

Double images.

Monocular and binocular vision.

Stereoscopic effects.

Irradiation.

Angular rotation.

Region of distinct vision.

The yellow spot.

The ophthalmoscope.

Tests for color blindness.

XIV. Hearing.

Anatomical.

Comparative: The ear in *invertebrates* is a sac provided with modified (hair) cells, and enclosing otoliths and fluid; in the highest *mollusks* (cephalopods) there is a membranous and cartilaginous labyrinth.

Vertebrates: most *fishes* have a *utricle* communicating with the semicircular canals.

Most *amphibia* have no membrana tympani.

The frog has a membrana tympani which is connected by three ossicles with the *Fenestra ovalis.*

Reptiles have an *appendix of the saccule* corresponding to the cochlea; a fenestra rotunda is present.

In *crocodiles* and *birds* the cochlea is divided into a *scala tympani* and *scala vestibuli.*

Snakes have no tympanum.

In *birds* and *reptiles* the ossicles are represented by one bone, the *columella.*

Highly movable external ear of mammals.

The human ear:

The *external, middle* and *internal* ear.

Meatus: external and internal. Ceruminous glands and hairs. *Tympanum:* an air chamber; contains *vibrating* mechanism: Membrana tympani and ossicles (malleus, incus, stapes).

Communication with pharynx by the *Eustachian* tube.

Membrana tympani: shaped like a shallow funnel; the *umbo.*

Muscles: tensor tympani, laxator tympani, stapedius.

Handle of malleus attached to membrana.

Internal ear (Labyrinth). A bony labyrinth having *perilymph* around a membranous labyrinth enclosing *endolymph.*

Bony labyrinth consists of: Vestibule, semicircular canals and cochlea.

The *fenestra rotunda* is situated in the bony wall between tympanum and *scala tympani;* the *fenestra ovalis* in the same, opening into the vestbule.

The system of spaces filled by *endolymph* is the only part containing auditory *end-organs.*

These spaces all communicate with one another: the *semicircular canals* with the *utricle* directly; the *ductus cochlearis* with the *saccule* through the canalis reuniens; and the *saccule and utricle* by the *succus endolymphaticus.*

There are *auditory hairs* on the *maculæ* of the vestibule, and on the *cristæ* of the *ampullæ* of the semicircular canals.

Otoliths within the vestibule and ampullæ.

The spiral cochlea; divisions and connections.

Minute anatomy of the cochlea in special detail.

Physiological.

The external ear in man; its importance in animals with large movable ears.

Membrana tympani.

Has no fundamental note of its own.

"Damping" effect of the ossicles.

Auditory ossicles.

Vibrations are conveyed by the ossicles to the *endolymph* of the labyrinth with *diminished amplitude* but *increased intensity.*

The ossicles vibrate *en masse.*

Muscles:

Tensor tympani: by its contraction acts as a damper; when quiescent prevents the membrane from being pushed out unduly.

Laxator tympani: may act to draw the drum head outwards.

Stapedius: prevents too violent shocks being communicated to the endolymph through the stapes.

The action of these muscles is probably owing to nervous reflexes.

Eustachian tube:

Maintains equilibrium of pressure.

Is open during swallowing.

Abnormal:

Rupture of membrane tympani; thickened and indrawn membrana; catarrh and suppurative inflammation of the tympanum; continuous closure of the eustachian tube.

Physical: Sound viewed physically and physiologically.

Sound is due to vibrations of the air.

When these are *rhythmical* they produce a *note*.

A note has (*a*) Loudness dependent on the *amplitude* of the vibration. (*b*) **Pitch**, determined by the *rapidity* of the vibrations. (*c*) **Quality**, decided by the *over-tones* accompanying the fundamental note.

Harmony, discord, beats.

Sympathetic vibrations.

AUDITORY SENSATIONS.

The terminations of the auditory nerve are connected with the *maculæ* and *cristæ* and with the *basilar membrane*.

The *end-organs* are stimulated by the vibrations of the *endolymph*.

While this mechanism is intact deafness cannot be *absolute*.

Conduction of sound through the *bones* of the skull.

The organ of Corti may be concerned in the origination of impulses which give rise to the perception of relative pitch; but this is not demonstrated, and there are serious objections to this view.

The *basilar membrane* may, possibly, discharge this function.

Complex auditory sensations arise, it is likely, from the mingling of primary sensations.

Inaudible tones: variations with groups of animals, races of men, and individuals.

Entotic phenomena.

Auditory judgments:

"Projection" of sound into the external world.

We judge of the source, distance and direction of sounds by their loudness, pitch, quality, etc., but very imperfectly.

Sounds not previously heard are misleading; ventriloquism.

Abnormal: Deafness from brain disease.

XV. The Sense of Smell.

The essential olfactory end-organs are situated in the mucous membrane of the nose; the relative extent of the mucous membrane in different animals seems to bear some proportion to the acuteness of smell.

In man the olfactory organs proper are limited to the upper and middle *fossæ* of the nose.

The nose is lined with ciliated epithelium.

The olfactory cells.

Odorous particles reach the olfactory region by *diffusion* and by inhalation.

Odorous particles in liquids (solutions) cannot be smelled when in direct contact with the mucous membrane of the nose.

A certain period of stimulation precedes the development of the sensation, and the latter outlasts the stimulus.

The localization of odorous bodies is very imperfect.

Pungent substances excite the fifth nerve, but do not cause smell properly so-called.

Subjective sensations.

XVI. The Sense of Taste.

Taste-buds and other modifications of the epithelium.

The lingual (gustatory) nerve is distributed to the anterior part of the tongue, the glossopharyngeal to the posterior part and the pharynx.

Substances must be *dissolved* in order to be tasted (chemical stimulation).

Taste sensations may be generated by mechanical and electrical stimulation.

Tastes: Sweet, sour, saline, bitter, etc.

Some "tastes" are really smells.

Subjective tastes.

Abnormal: Deficiency of taste and smell often associated.

XVII. General Sensibility and Tactile Sensations.

Various forms of end-organs in the skin.

General sensibility.

Perfect localization of tactile sensations.

A tactile sensation is replaced by that of pain when the epidermis is removed.

The specialized sensations are those of touch, temperature and of muscular action.

All these sensations merge into pain when sufficiently intense.

PRESSURE SENSATIONS.

Weber's law applied to pressure sensations.

All parts of the skin not equally sensitive to pressure stimulation.

Variations of pressure more readily distinguished when successive than simultaneous.

A cold body is pronounced heavier than a warm one of equal weight.

Effect of contrast.

Tactual nerves or nerve fibres as distinct from other sensory fibres.

TACTILE SENSATIONS.

"Field of touch" composed of tactile areas.

Improvement in touch traceable to more exact limitation of the sensation areas in the brain.

Erroneous judgments: (*a*) From irritation of the nerves of the stump of an amputated limb. (*b*) Aristotle's experiment.

THE TEMPERATURE SENSE.

The temperature of bodies must be different from that of the surface of skin to which they are applied before they can give rise to any sensation of temperature.

The intensity of the sensation is in proportion to the rapidity of the change in the temperature of the skin as well as the actual temperature of the stimulating body.

The range of most accurate sensation is between 27 and 33° C., *i.e.*, near the normal temperature of the body.

The regions most sensitive to temperature do not correspond with those most sensitive to pressure.

There is probably a special nervous apparatus for each of the senses of pressure, touch and temperature.

THE MUSCULAR SENSE.

This sense is associated with muscles when in action, and to some extent when at rest.

It is of peripheral origin; but may be closely related to conscious volition.

Abnormal. Locomotor ataxy: inco-ordination related to loss of the muscular sense and tactile sensation.

Recent investigations on the temperate sense (Donaldson and others.)

EXPERIMENTS.

1. Determination of tactile *localization* capacity with bone, wooden or metal compasses (blunt or cork tipped).

2. *Temperature:* application of a blunt metallic point dipped into hot water (70°) to the same parts as in experiment 1.

3. *Muscular sense. Weight.*

(*a*) Determination of the smallest difference in weight appreciable when the weight is held in the hand.

(*b*) Repeat with the hand and arm supported on a table (sense of pressure).

4. Estimation of the *proportionate difference* appreciable. (*a*) When the weights are light, say, 1, 2, 3, 4, 5, grams; and (*b*) when heavier, say, 10, 20, 30 or 100, 200, etc., grams.

5. Plunging the finger into mercury and then gradually withdrawing it.

6. Placing two light weights, one cold and the other warm, on the tip of the finger, forehead, etc.

7. Selection of central spot in the palm of the hand, and application of a warm body and then of a light feather to this.

8. Placing the elbow first in warm water and then in a freezing mixture. There will be a sensation of pain in the fingers and of cold in the elbow.

9. Aristotle's experiment (tactile delusions).

Taste and smell.

1. Pinching the nose tightly, with the eyes shut, attempt to distinguish between pieces of apple, onion and potato.

2. Wipe the tongue entirely dry, and place on the tip a crystal of sugar and on the back part one of quinine, neither will be tasted until dissolved.

3. Apply non-polarizable electrodes to different parts of the tongue. The taste experienced is acid at the anode; alkaline at the kathode.

4. Place two pieces of sugar of equal size one on the tip and the other on the back of the tongue. The sensation will be most acute at the tip.

5. Place a drop of solution of quinine on the tip of the tongue; it will be but little tasted; place another drop on the back of the tongue, it will be readily tasted.

Sound and Hearing.

Sound from vibration; pitch; loudness; quality; overtones; resonance; harmony; discord, etc.

Estimation of the direction and distance of sounds.

NOTE.—Many of the above experiments are after Foster and Langley.

XVIII. The Voice and Speech.

Voice is produced by the vibration of the vocal bands (cords).

Anatomical and physiological.

The laryngeal mechanism consists of: A skeleton composed of cartilages, held together by ligaments, permitting a certain degree of mobility of the cartilages on each other.

These movements are effected by muscles; and their purpose is the regulation of the *width* of the respiratory opening (glottis) and the degree of *approximation* and *tension* of the vocal bands.

The cartilages: The arytenoid cartilages are of especial importance in voice production.

The processus vocalis.

The attachment of the *vocal bands.*

The superior vocal bands (*false cords*).

Ventricle and *sacculus laryngis.*

The *rima glottidis.*

The mucous membrane of the larynx.

The muscles grouped in antagonistic pairs.

Their attachments.

The "sphincter" of the larynx includes the:

Thyro-ary-epigloticus.

Thyro-arytenoideus externus.
Thyro-arytenoideus internus.
Arytenoideus posticus.

Collectively and singly, except the last, they tend to close the glottis, sphincter-like.

Mechanism for *widening* the glottis; *tightening* the vocal bands; *slackening* the vocal bands.

Nerve supply of the vocal mechanism.

The superior laryngeal supplies the mucous membrane and the *crico-thyroid.*

All the other muscles are supplied by the recurrent (inferior) laryngeal.

Effects of division of the laryngeal nerves.

The utterance of a note involves complex co-ordination.

The requirements for the utterances of a note are:

1. A certain degree of tension of the vocal bands.
2. Approximation of their edges, which must be free from irregularities.
3. An expiratory blast of air.

Resonance chambers.

The quality of the voice is dependent chiefly on the influence of parts above the vocal bands.

Vibration of the vocal bands.

The registers of the voice.

The chest voice; falsetto voice.

In the *chest* voice the vocal bands vibrate throughout their whole breadth; in the *falsetto* only in a portion of their breadth.

The range of human voices.

Male voices classed as: bass, baritone, tenor. Female as: contralto and soprano.

Recent investigation: The pitch of the voice; the registers; the falsetto voice (Mills).

SPEECH.

Speech as distinct from *voice.*

The vowels: each has its own characteristic disposition of the parts concerned in modifying the note produced by the vocal bands.

Illustrative examples.

Classification of consonants.

1. According to the place at which the modification takes place as *labials, dentals, gutturals.*

2. According to the character of the movement giving rise to them as *explosives, asperates, resonants* or *nasals.*

Whispering is speech without the use of the vocal bands.

Abnormal: Tumours on the vocal bands; paralysis of the muscles of the larynx.

DEMONSTRATION.

The laryngoscopic demonstration of the glottis. (*a*) In respiration. (*b*) In phonation.

XIX. Locomotor Mechanisms.

The movements of the limbs are effected by the muscles acting on them according to the mechanical principles governing levers.

Lever of first order P̄ F W̄; nodding the head.

Lever of second order P̄ W F̄; raising body on toes by muscles of the calf.

Lever of third order W̄ P F̄; raising forearm by biceps.

Examples of the third kind are most common; of the first rare.

Various kinds of joints.

In the *erect posture* the line of gravity must fall within the area of the feet.

The line of gravity of the head falls in front of the occipital articulation.

The centre of gravity for the combined head and trunk lies at about the level of the ensiform cartilage; the line of gravity of the whole body passes in front of a line drawn between the ankle joints.

The weight of the whole body is therefore sustained by the arch of the instep.

The mechanics of walking; standing on one foot; running; jumping.

The locomotion of quadrupeds.

XX. Reproduction and Development.

The hypotheses by which existing organisms are accounted for are:

1. *Abiogenesis* (spontaneous generation): living matter derivable from non-living matter.
2. *Special creation* of each principal form.
3. *Evolution*: "Evolution is a progress from an indefinite, incoherent homogeneity to a definite, coherent heterogeneity, accompanying an integration of matter and dissipation of motion" (Spencer).

Organic evolution relates to the origin of organized bodies.

The principles of organic evolution:

(*a*) All forms of life have been derived from one or a few antecedent forms.

(*b*) There is in organisms a tendency to vary in form.

(*c*) There is in organisms a tendency to inheritance of ancestral forms ("Heredity").

(*d*) By the principle of "natural selection" ("survival of the fittest") certain forms survive, others perish.

Origin and extinction of species and varieties.

The following grow out of the above:

(*a*) Excess of development of some parts relating to others.

(*b*) Complete or partial suppression of some parts.

(*c*) Coalescence of parts originally distinct (after Huxley).

Historical:

The doctrine of organic evolution did not *originate* with Charles Darwin but was previously independently worked out with greater or less fullness by different scientists.

Charles Darwin was antedated slightly by Wallace in the promulgation of the most complete form of the doctrine of organic evolution.

The part of the doctrine of organic evolution really *originated* by Darwin was "natural selection"; "descent with modification" had been previously announced.

"Organic evolution" is regarded by the great majority of scientific men as a probable *hypothesis*.

Objections to the hypothesis by Darwin himself and others.

REPRODUCTION IN THE LOWEST ORGANISMS.

1. *Fission* in the protozoa, and many unicellular plants.

2. *Gemmation* (budding) in infusorians, cœlenterates, torula.

3. *Asexual conjugation* in infusorians, fungi.

4. *Metamorphosis* in insects.

5. *Alternation of generations:* In tœnia solium, tœnia medio-canellata, cysticercus cellulosæ, tœnia echinococcus.

6. *Parthenogenesis* in insects, crustaceans.

7. *Sexual reproduction* (without intermediate stages) in most vertebrates.

The fundamental biological law of Haeckel: "The ontogeny is a short repetition of the phylogeny."

REPRODUCTION IN VERTEBRATES.

"*Omne animal ex ovo.*"

The ovum of the female and the spermatozoon of the male are cells.

The new organism results from a union of these cells.

In most cases this is effected by sexual conjugation; but this is not essential and does not take place in some large groups of animals.

PRODUCTION OF OVA AND SPERMATOZOA, AND CONJUGATION.

Semen consists of spermatozoa mixed with the secretions of the glands of the *vas deferens, Cowper's glands,* of the *prostate* and of the *vesiculæ seminales.*

Spermatozoa develop by the fission of the cells of the seminal tubes.

Chemical composition of semen.

Anatomy of the male generative organs.

The *ovum* is the product of a *Graafian follicle.*

Structure of the ovum and the Graafian follicle.

The G. follicles are derived from the *germinal* epithelium, which latter migrating inwards forms the *ovarian tubes* ("egg tubes") of the ovary.

Holoblastic ova (frog, mammal); *meroblastic ova* (birds, monotremes among mammals, reptiles, fishes, except the cyclostomata).

Chemical composition of the ovum.

Sexual maturity is the period at which perfect ova and spermatozoa are produced.

In the human species this period is called *puberty*.

Moral, mental and physical changes at puberty.

The age at which puberty is established.

The menopause (grand climacteric) of women.

OVULATION AND MENSURATION.

These are generally associated.

The menstruation of the human female corresponds with the "heat" or "rut" of lower mammals.

The characteristic accompaniments of menstruation are:

Changes in the uterine mucous membrane and contiguous structures especially the ovary. The rupture of the Graafian follicle may precede, accompany or follow the menstrual flow.

The *corpus luteum of menstruation* as distinguished from the *corpus luteum of pregnancy*.

The part played by continued increased vascular supply and by fatty degeneration.

Erection of the penis.

The peculiar structure of the *corpora cavernosa* and *corpus spongiosum* of the penis.

Erection is due to increased vascular flow by the dilation of the arteries of the penis (*anteria profunda penis*) excited by impulses passing along the *nervi erigentes* (dog) which arise from the second (or third) sacral nerves; the blood within the organ is retained by compression of the veins (by deep *transversus perinei*, and *ischio-cavernosus* muscles).

Erection is a reflex act the *centre* being in the lumbar cord (Goltz).

The centre may be stimulated or controlled by special psychical activity of the cerebrum.

Emission (ejaculation) of semen is effected by peristaltic action of the *vasa deferentia* and *vesiculæ seminales*; and rhythmical action of the *bulbo-cavernosus* muscle. The centre is in the lumbar cord.

IMPREGNATION.

During copulation there is said to be a peristalsis of the fallopian tube towards the uterus.

The ovum in the human subject is probably most frequently impregnated in the *fallopian* tube.

By the *ciliated* epithelium of the tube the ovum is conveyed into the uterus.

EARLY STAGES IN THE DEVELOPMENT OF OVA.

Spermatozoa enter the ovum by the *micropyle* (when present).

Within the ovum the cilium of the male cell disappears, the head becoming now the *male pronucleus*, and uniting with the *female pronucleus*, forms the first *segmentation* (cleavage) *nucleus*.

The germinal vesicle throws off a *polar globule;* the remaining portion constitutes the *female pronucleus*.

Segmentation of the ovum may take place without access of spermatozoa (parthenogenesis).

Partial parthenogenetic segmentation in the hen's egg.

Kinds of segmentation; the *morula*; the *gastrula*.

The arrangement of cells forming the *blastoderm*.

Originally the blastoderm consists of two layers: *epiblast* and *hypoblast*. Very soon the *mesoblast* is interposed.

The *area pellucida* and *area opaca*.

The *primitive streak* and *medullary groove*.

The *medullary folds*.

Laminæ dorsales; laminæ ventrales; chorda dorsalis or *notochord*.

The *protovertebræ*.

Head and tail ends of the embryo; flexures of these.

Formation of the membranes exemplified by diagrams.

The *amnion* and *yolk sac* of the chick are represented by the *placenta* of the mammal.

True and false amnion.

The amniotic cavity; gradual diminution of the yolk sac.

CHANGES IN THE HUMAN SUBJECT.

The impregnated ovum rests on the highly vascular *decidua vera* of the uterus, which at this place becomes the *decidua serotina;* part of the latter grows up around each side of the ovum as the *decidua reflexa*.

The ovum has its own proper coverings:

The shaggy chorion outermost; the amnion innermost.

The place of junction of the decidua serotina and proper membranes of the ovum marks the site of the placenta.

The *placentation of mammals:*

1. *Mammalia implacentalia* (monotremata, marsupialia).

2. *Mammalia placentalia.*

The second class is divided into *non-deciduate* (pachydermata, cetacea, solidungula, camelidae) and *deciduate.* The latter are *zonary* (carnivora, pinnipedia, elephant) and *discoid* (primates, insectivora, edentata, rodentia).

The placenta is *diffuse* in the horse, etc.; *cotyledonary* in ruminants.

Minute structure of the placenta in the human subject.

Sinuses, villi, tortuous vessels, etc.

The development of the embryo.

Structures formed from each layer of cells of the blastoderm.

The neural canal; protovertebræ.

The notochord is finally represented by the centres of the bodies of the vertebræ.

Gradual additions to the original epithelium of the medullary folds, by which the spinal cord is formed.

The brain: Original three vesicles; subdivision of the hindmost; outgrowths and thickenings.

Beginnings of certain organs of special sense as vesicles and pits of the brain.

The *heart:* First a straight tube; later twisting and subdivisions.

Relations to the permanent form of the heart in lower groups of animals.

The circulation as first established (vitelline).

The later circulation in the chick.

The fœtal circulation; the umbilical cord has *two arteries* and one vein.

The best blood is distributed to the head and upper extremities of the fœtus.

Branchial arches.

Remains of structures once functional:

Umbilical arteries remain as *lateral ligaments* of the bladder; parts remain as the *superior vesical arteries.*

Umbilical vein remains as the *ligamentum teres.*

Ductus arteriosus as the *ligamentum arteriosum.*

Development of the *alimentary canal* from an unclosed tube of mesoblast lined by hypoblast.

Development of the respiratory system from the outgrowths of the alimentary tract.

The *bladder* and *urachus* are the remnants of the allantois.

The development of the eye in detail.

Development of the genito-urinary system.

Müllerian Ducts.
(Ducts of the Pronephros).

Female.	*Male.*
Fallopian tubes.	Hydatid of Morgagni.
Hydatid.	Male uterus.
Uterus and vagina.	

Wolffian Bodies (Mesonephros).

Parovarium.	Vasa afferentia, Coni vasculosi.
Paroophoron.	Organ of Giraldès, vasa aberrantia.
Round ligament of the uterus.	Gubernaculum testis.

Wolffian Ducts.

Chief tube of parovarium.	Convoluted tube of epididymis.
Ducts of Gaertner.	Vas deferens and vesiculæ seminales.

Metanephros.
Kidney.
Ureter.

(Stirling after Quain.)

THE PHYSIOLOGY OF DEVELOPMENT.

THE NUTRITION OF THE EMBRYO.

Inasmuch as the ova of many groups of animals pass through their entire development in water, they must contain within themselves the energy (latent) which becomes manifest in the processes of development.

The yolk of the bird's egg furnishes the nutriment for the development of the embryo.

Oxygen passes through the shell to the tissues either by the allantoic vessels or directly.

Prior to the development of the placenta the mammalian embryo is nourished by *diffusion*.

The placenta has *nutritive, respiratory* and *excretory* functions.

The same laws as govern the tension of gases, etc., in the respiration of the adult apply in the case of the interchanges between the maternal and fœtal blood.

Fœtal blood contains relatively less hæmoglobin.

Tissues of the fœtus composed at first of but little differentiated protoplasm; later they abound in glycogen; this probably serves as the crude material for the developmental metabolism.

Digestion is in abeyance during fœtal life; the liver is functionally active, probably also the skin and kidneys.

Meconium of the alimentary canal.

The amniotic and allantoic fluids may contain urea and allantoin respectively.

The changes taking place in the circulation after birth are owing to changes in *blood pressure* induced by alteration in the environment of the animal.

The work of the right vetricle is greater than that of the left in the fœtus.

The first respirations of the infant are due to stimulation of the respiratory centre by blood deficient in oxygen, together with the influence of afferent impulses from the skin.

PARTURITION.

Variations in the period of gestation of different groups of animals.

In woman it lasts about 280 days.

Mechanism of parturition.

Parturition is a reflex act; the centre is in the lumbar cord (Goltz).

The expulsion of the placenta; after-contraction of the uterus preventing hæmorrhage; gradual fatty degeneration and absorption ("involution") of the superfluous uterine muscle cells.

XXI. The Phases of Life.

At birth the child is nearly $\frac{1}{3}$ the maximum length and $\frac{1}{20}$ the maximum weight.

Relative proportion of parts.

Rapid increase in stature and weight in the earlier years.

The *digestive system*: Saliva, gastric juice, etc., active.

The *circulatory and respiratory system*: Heart relatively larger; circulation and respiration more rapid, which points to rapidity of metabolism.

The absorption of O is more rapid than the production of CO_2; constructive metabolism.

The *temperature* is slightly higher than that of the adult (0.3).

More *urine* is passed relatively to body weight; the quantity of urea, etc., is greater; of phosphates less.

Lymphatic system is prominent.

Large thymus and thyroid glands.

The brain is large and the nervous system is more irritable.

Stimulation of the cortex is not followed (in newly born animals) by the usual localized movements.

The senses in the infant.

The dentition of the young child; temporary or milk teeth; permanent teeth.

The changes in both sexes at puberty.

Woman and man compared.

The declining man; the old man: weakness and general rigidity; calcareous and fatty degeneration with age.

The effects of the seasons.

Hibernation.

Sleep:

Sleep is accompanied by loss of consciousness and a lowering of the rate of activity of all those processes which are ceaseless.

Condition of the eyes.

Death:

The object of an individual animal's existence is the reproduction of itself in offspring.

Death may result from gradual decline and final rupture at the weakest point; or by violent disturbances of normal processes, as in disease.

Death is by heart, lungs or brain (Bichat).

Somatic death and tissue death.

Demonstrations.

The human placenta and membranes.
The same parts of other animals.
The fœtus of different animals.
Microscopic sections of the embryo chick, etc., etc.

APPENDIX.

CLASS LABORATORY EXERCISES.*

FOOD-STUFFS, ARTIFICIAL DIGESTION, ANIMAL LIQUIDS.

I. Starch, Dextrin, Dextrose, Fats.

$C_6H_{10}O_5-$

1. **STARCH.** Insoluble in cold water; dissolves imperfectly to an opalescent liquid with heat.
2. The *cold* solution gives a blue color with solution of iodine.
3. Heat gently in a test tube till the color *begins* to disappear, then immerse the tube in cold water; the color will return.
4. **DEXTRIN.** Soluble in water; the solution gives a red brown color with solution of Iodine.
5. Treat as starch (in 3); the color vanishes but does *not* reappear on cooling.

$C_6H_{12}O_6$

6. **DEXTROSE.** *(Grape sugar)* crystalline, readily soluble in water, less so than cane sugar; reduces metallic oxides.
7. *Tests.* (a) Trommer's: To a couple of drops of solution (10%) of Cu S O₄ add K O H (or Na O H) till a clear blue solution results, then add to this a small quantity of the

---

* These exercises are based on those, contained in the Practical Physiology of Prof. Burdon Sanderson, which have been in use, under the author's direction, by students of McGill University during the past five years, and have proved eminently suitable.

Directions for individual work and for experimental physiology will be otherwise provided.

solution of sugar and heat gradually to bo...g. Boil for half a minute and let stand. Note the various changes.

Reactions: (1) $CuSO_4 + 2(KO...) \ldots CuO + H_2O + K_2SO_4$.

(...) CuO is *reduced* by the sugar to Cu_2O.

...light yellow that first appears is the hydrated Cu_2O.

Moore's test: Heat a solution of sugar with solution of KOH, ...lor changes to a shade of brown, the depth depending on the strength of the saccharine solution.

9. *Bismuth* (Böttger's test): Heat a solution of sugar with a pinch of bismuth subnitrate; a brown color and dark precipitation (on standing) results.

10. *Fermentation test*: A solution of sugar with a little yeast added placed in a test tube and kept in a beaker of water at about 17° C gives off CO_2 and forms alcohol.

Reactions: $C_6H_{12}O_6 = 2(C_2H_5OH) + 2CO_2$.

11. *Conversion of starch into dextrose:* $C_6H_{12}O_6 + \ldots$

To a dilute solution of starch, add a few drops of dilute H_2SO_4 and boil for 5–10 minutes.

The starch solution becomes limpid.

Test for sugar by Trommer's method.

The liquid contains also dextrin and unaltered starch.

12. *Fats.* Insoluble in water; soluble in ether, hot alcohol, chloroform, benzol, carbon bisulphide, turpentine, etc.

13. Boil a little butter in a solution of KOH in a beaker or porcelain capsule, for some time, a soap is formed; test by shaking up with soft water. —

To some of the potash soap add a little $CaCl_2$ in solution; a calcium soap (insoluble) is precipitated. Repeat with $MgSO_4$.

14. Add to the potash soap in a test tube a little strong

H_2SO_4 and boi[...] standing a layer of fatty acid rises to the top; it smells of butter.

Tests: (a) Pou[...]r paper a little ethereal solution of fat; on the evap[...] of the ether, a characteristic stain remains.

(b) Pour a little ether over the suspected liquid (e.g. s[...] solution); rapidly filter into a clean dry watch glass or [...] tube; evaporate; if fat is present it will remain [on the] glass. This test applies also to fatty acids.

Reactions:

| Tri-olein. | Potassium Hydrate. | Potassium tri-oleate. | Glycerine. |

$$3\,(C_{16}H_{33}O_2) + 3\,KOH = 3\,(C_{16}^-H_{33}^-O_2^-) + C_3H_5(OH)_3$$
$$C_3H_5. \qquad\qquad\qquad K_3$$

$$2\left\{\begin{array}{c}3\,(C_{14}H_{33}O_2)\\K_3\end{array}\right\} + 3\,H_2SO_4 = 6\,(C_{14}H_{34}O_2) + 3\,K_2SO_4$$

Oleic acid.

Inference from the experiments of exercise I:

Starch is insoluble in cold water; imperfectly soluble in hot water; gives a blue color-reaction with iodine; may be converted into dextrose.

Dextrin is soluble in water; gives a red color-reaction with iodine

Dextrose is soluble in water; reduces metallic oxides.

Fats are compounds of glycerine and a fatty acid; may be decomposed by a strong mineral acid; can form soaps; are insoluble in water, etc.

II. Foodstuffs.

MILK, FLOUR, BREAD.

These furnish specimens of all the classes of foods, organic and inorganic.

1. **MILK.** Reaction alkaline. Specific gravity 1025 to 1030.

Skimmed milk has a *higher* S. G.

(a) **PROTEIDS.** Place about a wine-glassful of milk in a flask, add a few drops of dilute acetic or sulphuric (25%) *acid*; gently warm to a little above blood heat; a granular precipitation of *casein* (with fat) will take place. Let stand.

(b) Treat in a similar way with a few drops of *extract* of *rennet*; a gelatinous firm clot forms. Let stand.

(c) Filter and test the filtrate (whey) for milk sugar (Trommer's test).

The salts cannot be conveniently tested for on account of the presence of proteid matter.

(d) Test the coagulated casein remaining on the filter for entangled fat (butter) as in chapter I, 14.

2. **FLOUR.** Wash a dessert-spoonful of flour through a piece of fine muslin held as a bag.

Test for starch and sugar (little or none).

The remnant of the washing is *gluten*. It is very tenacious. It gives proteid reactions (Millon's reagent).

3. **BREAD.** Digest in a capsule with warm water. Filter through paper.

Test filtrate for starch and sugar.

Test the residue for proteid and starch.

Conclusions: *Milk* contains all the essentials of a food stuff (fats, carbohydrates, proteids and salts); the "particulate" casein and globulin may be precipitated by chemical treatment; the salts are in solution in the water of milk.

Flour contains starch and proteid matter.

Bread, owing to the heat used in cooking, has had part its original starch converted into dextrose.

III. Proteids: Albuminous Substances.

Preparation of Albumin: Cut up white of egg with scissors; dilute 8-10 times with water; filter through linen and afterwards through paper.

TESTS FOR PROTEIDS.

1. Boil about a teaspoonful of the solution in a test tube; it coagulates.

2. Treat the *coagulum* with strong mineral acids (HCl, H_2SO_4); it does not dissolve.

3. Add to a similar portion strong HNO_3. It coagulates and on boiling turns yellow.

4. Cool the test tube (3) and add strong NH_3. The precipitate becomes orange (*Xanthoproteic reaction*).

5. Add to another portion a drop of Millon's reagent. The resulting precipitate turns red on boiling. Millon's reagent is Hg dissolved in its own weight of strong HNO_3 and then diluted with twice its volume of water.

6. Add to another portion Potassium Ferrocyanide (K_4FeCN_6) and acetic acid (CH_3COOH.) A white precipitate falls.

7. To another portion add a drop of solution of $CuSO_4$ and a little of a solution of KOH. A violet color appears.

8. **ACID-ALBUMIN.** Add to a dessert-spoonful of the prepared solution of albumin a few drops of diluted acid (about (5%); gently warm for some minutes; *acid-albumin* is formed; ascertain that it does not coagulate on boiling by testing a separate portion; then divide into two parts. To one part

www.ingramcontent.com/pod-product-compliance
Lightning Source LLC
Chambersburg PA
CBHW032148160426
43197CB00008B/815